THE ENERGY CENTER:
New Alternative for Effective Energy Use

THE ENERGY CENTER:
New Alternative for Effective Energy Use

by

JOHN F. HEMDAL, P.E.

Research Engineer
Environmental Research Institute of Michigan

and

Private Consultant
Ann Arbor, Michigan

FOREWORD

The ready availability of inexpensive energy in an easily utilized form is a necessity in a modern technological economy. A rapid decline of energy availability would burden society as we know it with hardship, depression and, probably, social chaos. But how are we to maintain the flow of energy? Should we put all our resources and effort behind a switch to nuclear power or to a total coal economy? Shall we drop all other research and development in order to pursue solar energy? Shall we take up the so-called "Soft Path" approach and emphasize conservation? A mistake at this point could be eventually disastrous.

It is the author's opinion that the entire population can and should contribute to the determination of our energy future—that it should be an effort of the collected wisdom of an involved and *informed* public. The purpose of this book, then, is to provide sufficient information, at least for one alternative, to assist planners, architects, engineers and government officials, as well as the general concerned reader, in this decision-making.

How will consumer and industrial energy needs be successfully and efficiently met in 1985? 1990? 2000 and beyond? Waiting until the future arrives before considering these questions is increasingly recognized as a dangerous gamble. Clearly vital energy decisions are immediate concerns. This book examines the "Energy Center" concept, a dramatic new approach to energy planning and utilization that may hold many practical answers for the future.

What is an energy center? How is it designed? How will it work? In what ways will it affect the lives of people? Will it prove environmentally harmful or benign? To what extent does the energy center promise energy efficiency and conservation more reliably than other options? These and related topics are technically and thoroughly analyzed.

The actual and potential shortage of energy in our society has been convincingly documented. It is not the purpose of this discussion to gauge the severity of this shortage, but rather to make clear that energy programs requiring a decade or more to plan and construct must be decided on now. The future will be too late.

In this book the energy center as a dynamic vehicle for energy efficiency is comprehensively explored. Energy centers obviously will require extraordinary foresight, planning, coordination, compromise and, perhaps, sacrifice. That is, the stark dimensions of the challenge are unmistakably implied: the energy center is a large-scale integrated complex containing both the sources of energy (the public or private utility) and the users of energy (*e.g.,* the factory). Major transportation lanes would also converge there.

In analyzing this challenge, general features of an energy center will be given and supplemented by specific examples from numerous contemporary studies. Our basic definition of an energy center emphasizes efficient use of primary fuels. Energy efficiency is the ultimate key to effective energy utilization and conservation. We can assume energy centers will generally be large and complex because energy efficiency for effective conservation strongly implies close coordination among energy generation units and energy users. Such coordination requires a diverse, interlocking network in which all energy sources from primary fuels to waste heat are fully utilized. A practical, continuous partnership of energies and resources is one expected dividend of the energy center approach.

Thus an energy center will be a large-scale energy generation complex with associated energy utilization elements coordinated and interlocked to derive the total heat value of fuels while using raw materials with optimum efficiency.

Will energy centers be bureaucratic and environmental nightmares or ultimate accomplishments in progressive energy technology? This book considers the problems and shows how pitfalls can be avoided. The position is taken that every realistic energy center plan must include conservation of the environment and natural resources as indispensable criteria. Inherent in energy efficiency is environmental conservation and the saving of natural resources. Consequently these considerations are fundamental aspects of the energy center concept.

The primary goal motivating construction and operation of an energy center is to meet the expanding regional demand for electric power generation. Energy centers will be designed to specifications meeting the particular needs of different regions. An energy center may or may not use coal or nuclear fuel as a primary source. It may or may not include a satellite city as part of its definition. It may or may not contain an agricultural complex within its range of influence. Local conditions will determine the design, which means that energy center planning cannot be expected to fit snugly in a standard pattern.

An energy center will contain, as minimum elements, large electric power generating units, collocated industrial facilities to use the resultant power, and a close cooperation among all units to achieve energy and resource efficiency and conservation. The exact configuration of an energy center will depend on the requirements of a given region, the availability of primary fuels, transportation modes, and environmental as well as social implications.

In this book a comparative study is made of energy centers conceived for various locations in the United States, including the following:

1. The Michigan Energy Center
2. Camp Gruber, Oklahoma, Energy Center
3. The Commonwealth of Pennsylvania Energy Park
4. Puerto Rico Energy Center
5. Wasatch Front Area, Utah, Industrial Complex

This section, from energy studies now in progress, provides the contrasts that explain the energy center concept and describe important variations as they appear in different regions. The Michigan Energy Center study will be covered in greatest detail. Comparisons will be made and differences illustrated by detailed reference to the other energy centers considered.

Fundamental factors determining the feasibility of an energy center in a given location are analytically assessed, including regional electric energy demands, land requirements, water availability, characteristics of suggested site locations, environmental and social impacts, economic benefits, transportation conditions and fuel availability. Throughout this study, the energy center as a pragmatic energy option is methodically dissected in terms of specific technical, social and environmental needs.

The energy center appears to be an idea whose time has come. It is an energy alternative that current technological and resource facts-of-life argue we should consider in depth and expeditiously implement. The concept contains an exciting promise of resource and energy efficiency combined with responsible environmental and social action. This book seeks to present the energy center concept objectively with balanced attention to beneficial as well as potentially negative features. Some portions may prove controversial, though hopefully these too will serve to illuminate a critically important subject that will absorb the attention of mankind for the remainder of this century and beyond.

ABOUT THE AUTHOR

Dr. John F. Hemdal, Registered Professional Engineer, is a Research Engineer with the Environmental Research Institute of Michigan (ERIM) and Adjunct Assistant Professor of Engineering at the University of Michigan, Dearborn Campus. He has an extensive private consultation practice, specializing in Energy Systems and Noise Control. Formerly he was an Associate Engineer in the Applied Physics Laboratory at Johns Hopkins University and an Instructor in Electrical Engineering at Purdue University. In the 1970s, Dr. Hemdal has been Principal Investigator on a feasibility study of Energy Centers in Michigan. The present book benefits from current background data thus acquired.

Dr. Hemdal holds three degrees in Electrical Engineering from Purdue University, including the PhD. His published work in books and journals is wide and varied. He is a Senior Member of the Institute of Electrical and Electronics Engineers. Other professional affiliations include Tau Beta Pi, Eta Kappa Nu, Acoustical Society of America, Pattern Recognition Society, and the Institute of Noise Control Engineering.

The author brings the knowledge and experience of a distinguished research and consultation career to this important pioneering book on Energy Centers.

PART ONE
THE ENERGY CENTER CONCEPT

PART TWO
IMPACTS: THE ENVIRONMENTAL, SOCIAL AND ECONOMIC EFFECTS OF ENERGY CENTERS

THE ENERGY CENTER CONCEPT

The goals of a large-scale energy center are economies-of-scale, energy conservation, symbiosis and the concentration of environmental impacts. The benefits of collocation, as opposed to the alternative dispersed sites method, are described in Part One.

Energy centers capable of producing 20,000 to 40,000 MWe would provide not only electric power but also process steam and low-Btu gas to the collocated industries. The advantages of this approach, such as possible agricultural and aquacultural use of waste heat for soil warming, greenhouses, fish culture and so on, will be examined also.

ENERGY CENTER STUDIES

The concept of collocation, the clustering of electric power generation facilities around industrial users of that power, is not a new idea. However, it is only recently (1968) that systematic analysis of such complexes has been performed. Now several studies have been completed allowing a preliminary evaluation of the economical and environmental impact which can be expected from energy centers as well as a comparison with the dispersed siting approach, the method of electric power generation now commonly used.

The following are definitions of energy centers used by investigators of the various studies. The definitions are similar and share several concepts. These common elements should be noted as key features of energy centers. Considering these common elements, we conclude that energy centers are large complex integrated power systems (mostly electrical and thermal) involving both generation and usage to benefit from economies-of-scale and symbiotic relationships. In this way the "whole may be greater than the sum of its parts" when the full range of economies and efficiencies is taken into account.

ENERGY CENTER DEFINITIONS

The Michigan Energy Center[1]: "An energy center is defined as a centralized power-generating facility serving the power needs of a region, coupled with a variety of industrial and agricultural energy users. The object of the energy center is to make efficient and economical use of the total thermal and electrical energy obtained from the nuclear and/or fossil fuels employed by the power-generating facility."

Camp Gruber, Oklahoma, Energy Center[2]: "An energy center is defined as a group of industrial plants that produce, consume or transfer large quantities of energy. Numerous types of industrial plants could be built in an energy center in accordance with the energy needs of the surrounding markets. The concept of an energy center generally includes the intention to group large numbers of energy-generating facilities at a single location in order to gain several advantages. As a general rule, an energy center is defined as containing facilities that generate or utilize at least 10,000 megawatts electricity (MWe)."

The Commonwealth of Pennsylvania Energy Park[3] : "An 'energy park' is a large concentration of electrical generating capacity at one geographic site. This concept has been proposed as an alternative to the conventional or dispersed siting pattern for minimizing the cost and impact of electrical generation. As proposed for the Commonwealth of Pennsylvania, the park will be an agglomeration of power plants, both nuclear and coal-fired, with associated support facilities."

The Puerto Rico Energy Center[4] : "The energy center concept calls for the integration of a low-cost source of energy with industries that require large amounts of energy either in the form of electrical power or as thermal energy such as steam. As such, it is not an entirely new idea; many instances are known of energy-consuming industries located near low-cost sources of energy obtained from gas, coal or hydropower. What is somewhat new is the conscious systematic approach to creating a complex of industries around a nuclear energy source such that their combined loads will permit the economies of large-size power plants to be realized and to select those industries that will profit from the low-cost energy, from the interchange of products and the sharing of facilities. In short, an integrated complex formed to make the whole more efficient than the sum of the individual parts."

The Wasatch Front Area, Utah, Industrial Complex[5] : "The proposed complex . . . consists of a power plant, a desalted water plant and an industrial manufacturing complex in close enough proximity to make the transfer of utilities and the employment of common services economically viable."

In the following section alternate configurations are given for each of these energy centers as well as other energy center concepts now in the planning stage for various parts of the U.S. This should give useful background information for subsequent discussions.

ALTERNATE CONFIGURATIONS

The Michigan Energy Center

The primary objective of a Michigan energy center is to meet the state's electrical energy needs from 1985 to 2000. Additional benefits include advantages resulting from locating other types of energy production, conversion or utilization facilities at or near the center site. By "collocating these additional facilities at the site, a number of benefits are gained: assurance of a reliable supply of various forms of energy to important industries, conservation of energy, a reduction of capital investment for both industry and power-generation facilities and a minimization of environmental impact."

A major objective of energy center development is to achieve substantial economic benefits by carefully matching the system to the special needs and characteristics of the region. These benefits can be achieved through economies-of-scale resulting from mass production, common construction and integration of facilities. Reduction of construction costs may also be achieved by combining complementary individual functions; for example, by using a corridor for both electric power transmission and materials transportation, or by reducing the needed capacity of cooling towers as a result of alternative and beneficial uses of waste heat.

Because of increasing national emphasis on energy conservation, increased energy efficiency and conservation is another important objective. Energy can be saved through a better utilization of the total heat energy derived from the fuel source. Low-quality heat, from which most usable electrical energy has been obtained, can be employed effectively in some agricultural or industrial processes. Finally, there may be a reduced need for the transport of materials because of the collocation of complementary industries.

A number of social and economic advantages could also be derived from the development of a Michigan energy center. The expansion and diversification of the local or regional economy is one. Proper development of urban, agricultural and recreational areas can be another positive impact on living conditions in the region.

In addition, the center, if properly designed, could have a positive environmental impact. Although environmental damage would not be completely eliminated, it could be confined to a more limited area than if the same total industrial capacity were dispersed.

The energy center configuration selected for the Michigan study depended on matching these objectives with the state's predicted energy needs and the special characteristics of the state's resources and markets.

Most energy center plans provide for generating sufficient power to meet a region's needs to the year 1990 or 2000, exclusive of current plans for additional capacity. In order to estimate the future requirements for electric energy in Michigan, use was made of the load forecasts developed by Consumers Power Co. and Detroit Edison Co., the largest utilities supplying electric power to the state. The methods used by these companies have been reviewed by the Michigan Public Service Commission. These studies project a future compound annual growth rate of around 5%, significantly less than the historical rates of around 7%. These forecasts were used to estimate the energy center capacity required to meet the projected new demand, not including current planned expansion or the normal reserve maintained by the utilities. The estimated size of the center is based on the assumption that about 80% of new capacity would be located at the center. Under these assumptions the requirements can be met by an energy center with a total installed capacity of 23,125 MWe. Approximately 1,500 MWe of capacity would be added per year from 1987 through 1994 and 2,000 MWe of capacity would be added from 1994 through the year 2000. The Michigan configuration calls for eight coal-fired units totalling 6,225 MWe and 13 nuclear units totalling 16,900 MWe. Transmission-line requirements for the energy center could be handled by approximately 12 lines of 765-kV capacity feeding power from the site.

For the Michigan energy center concept, it is maintained that a collocated industrial center would offer a number of potential advantages to the state. Industries proposed for this center include a pulp and paper mill, petroleum refiner and a chemical complex, all of which would require substantial quantities of process steam which could be supplied by the center's steam-generating systems. It has been calculated that the savings from the combined generation of electrical energy and 5 million pounds per hour of process steam would be equivalent to 1.5 million barrels of oil per year. A coal gasification plant at the center would supply low-Btu gas as a substitute for increasingly scarce natural gas. Another type of industry suggested for the center is a mini-steel mill, which uses as its raw material the plentiful quantities of scrap steel and scrap iron that are by-products of Michigan industries. The mini-steel mill would melt the scrap in electric arc furnaces. Some finished products from this mill, such as reinforcing bars, could be used in the construction of the center itself.

A biocomplex has also been proposed for the Michigan center which would cover 960 acres with labor-intensive high-productivity agriculture. This would make use of otherwise wasted heat produced during electric power generation. Subsystems proposed in the biocomplex include fish culture, grain drying, waste treatment and greenhouse warming.

Other features suggested for the Michigan center are port facilities to handle part of the 30 million tons of coal and oil inputs to the center, and a city of 200,000 needed to house center employees and their families, as well as the industry, commerce and government personnel servicing the center.

The Michigan concept—including the energy and industrial centers and biocomplex—would occupy a land area of about 13,470 acres and employ an operating force of 8,300. Total land including the city would require about 19,870 acres. Table I lists the types and capacities of collocated industry and agriculture envisioned for the center.

Table I. Collocated Industry and Agriculture

Industry	Capacity
Paper Mill	2,000 ton/day
Chemical Complex	$240 million annual sales
Petroleum	250,000 bbl/day
Coal Gasification Plant	250 million ft^3/day of gas (300 Btu/ft^3)
Mini-Steel Mill	150,000 ton/yr
Biocomplex (Site 1)	
Catfish Culture	25 20-ac ponds
Greenhouse	300 ac (vegetables and flowers)
Grain Drying	65 ac (20 million bushel facility)
Waste Treatment	85 ac of algae ponds (10 mgd)

The Camp Gruber Energy Center

Although many potential systems were considered for possible inclusion in the Camp Gruber energy center, a specific configuration was not selected by the study team. However their report[2] does specify which systems would and would not be appropriate for Camp Gruber.

Through their analysis, the Camp Gruber study team concluded that an energy center could be visualized as requiring electrical generation of from 15 to 30,000 MWe by the end of the century. They conclude that "this amount of generation would probably be obtained from nuclear and coal-fired plants. The need for coal gasification and/or liquefaction plants is also apparent. Energy-related industrial plants might also be a part of an energy center. Since many of these can be located at a wide variety of locations, economic and other considerations will probably dictate whether they would be sited at an energy center . . . it has not been determined whether industrial buildup is compatible with a nuclear energy center."

In the Camp Gruber study it was suggested that the heavy demand for petroleum products indicates a future market for conversion products from petroleum substitutes, probably taking the form of coal gasification or liquefaction. They also suggest that increasing energy requirements will create a greater need for energy-related industries such as nuclear fuel processing and fabrication plants, power plant component manufacture, energy conversion equipment and so on.

The Camp Gruber report discusses a variety of possible electrical generation facilities which might be included at the site, such as fossil-fired, nuclear, hydroelectric, geothermal, solar and wind-powered. Analysis of the siting requirements for specific energy facilities revealed numerous systems which could be appropriately sited at Camp Gruber. These elements include:

Oil-fired power plants	Magnesium reduction plants
Coal-fired plants	Agriculture-aquaculture
Nuclear power plants	Uranium enrichment
Solar power plants	Uranium conversion
Wind power plants	Nuclear fuel fabrication
Chemical energy storage	Nuclear fuel reprocessing
Electrical energy storage	Surface waste storage
Mechanical energy storage	Petroleum refineries
Pumped storage facilities	Pipeline stations
Coal gasification	Power plant component manufacture
Coal liquefaction	Energy test sites
Aluminum reduction plants	Hydrogen gas storage

The analysis also indicated energy facilities inappropriate to Camp Gruber. These include:

Gas-fired power plants	Use of gas contrary to national policy
Hydroelectric power plants	No hydraulic energy available on site
Geothermal power plants	No goethermal energy available
Aquifer heating for energy	Satisfactory aquifers not available
Coal mining	No identified commercial coal seams
Uranium mining	No nearby sources of uranium ore
Deep storage of wastes	Unsatisfactory geologic conditions
Oil shale processing	No nearby source of oil shale

The Camp Gruber analysis did not include the possibility of a new city but relied on surrounding communities to handle the influx of workers and provide necessary services. The analysis included a study of the impacts of the energy center on these surrounding communities.

The Pennsylvania Energy Park

The Pennsylvania study team has provided a "reference concept" of a standard energy park.[3] Their specifications for the standard energy park include provisions for a 10,000-MWe base load generation, composed of ten 1,000-MWe unit plants, five nuclear and five coal-fired. The "base load" specification implies that the park is not expected to respond to peak load demands. The specification also assumes a rural site, away from any urbanized area greater than 25,000 people, not within a Commonwealth air quality control region. The specifications stipulate that waste heat is to be dissipated via natural draft wet cooling towers. The Pennsylvania configuration will employ devices for SO_2 removal, fly ash removal and tall stacks to satisfy emission and ambient standards. The development of the park will require ten years for the first coal facility and twelve years for the first nuclear facility to become operational. After that, one plant per year will become productive for the remainder of the development program. The completed site will require a minimum of fifteen square miles of land. However, space required by switching yards, cooling apparatus, fuel, ash, water, sludge storage and nuclear exclusion zones may lead to a total contiguous area larger than the fifteen square miles. Little provision has been made for secondary industrial development to occur.

The Puerto Rico Energy Center

The Puerto Rico energy center concept is not as large as the other centers since it is intended to service a much smaller population. Nevertheless, its plan proposes many similar features including a dual-purpose power plant, a petrochemical plant, aluminum and salt recovery plants, a waste treatment area, a demonstration farm and a deep harbor. It is not the intent of this center to supply electric power to an extended region nor to meet future increased demand. Rather the bulk of the 2,785 MWt, including 540 MWe, will be required at the collocated industrial and agricultural facilities. The suggested center would occupy an area of approximately 2,500 acres and would need about 4,000 skilled operating personnel and a construction force of 4,000-5,000. A newly emerging city is not anticipated, but substantial growth in the neighboring town of Aguirre will occur.

The Wasatch Front Industrial Complex

The objectives of the Utah study include the desire to form an industrial complex which would economically utilize available resources (notably salt) in a combined energy and chemical center. The primary elements of the proposed complex are a power plant, a desalting plant and an industrial manufacturing complex. The electric power generation capacity was estimated to meet the needs of the complex itself; no provisions were made to serve the power needs of the surrounding region, as in some other concepts. Thus only 310 MWe of electric power generation is included.

Though no city is planned, the concept does share many of those features with other centers, including interrelated and collocated industry and energy production. The principal idea behind the Wasatch complex is to extract sodium and magnesium chlorides from brine and combine them with crude oil to manufacture a wide variety of petrochemicals. These would include chlorine, propylene, ethylene dichloride, olefins, vinyl chloride, polyethylene and others. In this study, particular attention was paid to the economics of various configurations, *i.e.,* to costs and expected rates of return.

The Hanford Nuclear Energy Center[6]

The Hanford concept has several distinctions. First, it is one of the largest, expecting to produce over 40,000 MWe by the year 2000. Second, its power generation is to come solely from nuclear power units. And third, there has been no effort to include collocated and integrated industries at the site, although it was recognized that certain industries might be attracted to the vicinity. A major emphasis in the Hanford study was to determine the economics and feasibility of including uranium fuel fabrication, uranium enrichment and nuclear waste storage facilities on or near the site. One reason the Hanford site was selected was that in recent years the Atomic Energy Commission has used the Hanford Reservation in relation to the Liquid Metal Fast Breeder Reactor (LMFBR) program, the production of nuclear materials, waste management activities and life sciences and environmental R&D programs.

Other Energy Concepts

The AEC, FEA, ERDA and other government agencies and administrations, both past and current, have commissioned studies into the feasibility of constructing and operating various energy centers or parks to be sited at a number of locations throughout the country.

These additional centers and parks will not be described in detail in this book, but reference will be made to any of their outstanding differences, problems or interesting features.

In the remainder of the book the Michigan Energy Center will be described in detail. Features of the other centers will be described whenever they illustrate contrasts, differences or topics of special interest.

The objective of most energy center studies is to determine how well the collocated and integrated components compare with the custom of scattering facilities, the common practice in the past.

II

THE NEED FOR AN ENERGY CENTER

THE NEED TO SUPPLY PROJECTED
ELECTRIC POWER DEMANDS

A primary objective of many energy centers is to meet a state or region's projected demand for electrical power over the next 20 to 25 years. These power demand projections, then, are used to arrive at estimates of generating capacity needed and, hence, the eventual size of the energy center. An example of how such estimates are made is illustrated by the Michigan center.

Michigan's two largest power companies (Consumers Power and Detroit Edison) supply the load for most of the lower peninsula. The noncoincident peak load for the two utilities was about 11,000 MWe for the summer of 1975. The peak load of the entire state includes about 1,000 MWe for the upper peninsula companies and about 1,000 MWe for the other lower peninsula companies. Other companies in Michigan include the Indiana and Michigan Electric Co., municipals and the Rural Electrification Administration. Most of these companies not only generate power but also purchase power from other companies.

Two factors are considered when describing the power capability of a utility. Although the demand for electricity varies considerably throughout the day, and again by season, there is always some demand. This constant demand is referred to as the base load. The peak demand which must be met determines the total MWe capacity of the system even though these peaks draw power only for relatively short periods of time. The peak load is met by generating units which can be started and stopped on short notice. Since they are only needed for short intervals they are usually comprised of units requiring a comparatively low capital investment such as gas turbines. On the other hand, these units usually have much higher fuel costs. Base load plants, on the other hand, involve high capital invest-ments, but operate at low fuel costs.

The Consumers Power Company system has a total of about 10 generating plants with a nameplate rating of 5,580 MWe. Almost every type of power plant is represented in the Consumers Power system including hydro, nuclear, coal- and oil-fired plants and gas turbines. The hydroelectric capacity is about 130 MWe. The largest nuclear plant at Palisades has a potential for 800 MWe.

Detroit Edison and Consumers Power have created a Michigan Power Pool to exchange power by interconnected lines whenever necessary. This allows each system to build and operate larger, more efficient units because the power is shared over a larger region. The following hypothetical situation illustrates how savings are possible with a larger system. Suppose both utilities need an expanded capability of 1,000-MWe capacity, say 500-MWe base load and 500 MWe peak load, for a total of 2,000 MWe for the two companies. Since larger units are generally more efficient and, therefore, more economical, it would be advantageous for one company to build and operate a 1,000-MWe base load unit and the other to build and operate one 1,000-MWe peak load capability. Then, each utility would share (more properly, sell) about half of its peak or base load as required. This arrangement is superior to the alternative of having each utility construct and operate a 500-MWe base load and 500-MWe peak load capability. There may also be another advantage of requiring less total capacity, since it is unlikely that the two utilities would need to meet either maximum base load or peak load at the same time.

There is a limit to the size of an individual generating unit, however, based on considerations of system security and the law of diminishing returns. Aside from the fact that generating efficiency does not increase with size beyond a certain point, no unit can be such a large portion of the total capacity that a routine or forced shutdown of the unit would place undue burden on the remainder of the system. Each company attempts to maintain a 20% reserve capacity which can be a shared responsibility between companies. However, the power system must maintain spinning reserve large enough to handle the load if its largest unit goes out. One way to maintain this reserve is to operate large base load plants at less than maximum load. This part-load point is surprisingly more efficient than maximum load.

An interesting and valuable approach to meeting peak demand is by use of a pumped storage facility. A pumped storage facility has reversible turbines which pump water to a higher level during periods of low demand. Then during periods of peak demand the water flows in reverse, generating electricity with the turbines running backwards. Consumers

Power operates a pumped storage plant at Ludington on Lake Michigan (51% owned by Consumers Power and 49% owned by Detroit Edison. This hydro storage facility provides peak power during the day and is refilled nights and weekends. The system has a yearly average efficiency of 71%. Its economic justification comes from permitting increased use of base plant generating power at 8 to 10 mils to save power from peak load plants costing 30 mils. It would be interesting to calculate the economics of installing wind-powered water pumps on the windy shores of Lake Michigan to keep the pumped storage reservoir filled.

Since the main objective of an energy center is to meet the power needs (primarily the base load requirements) of a region, it is necessary to describe load forecasting procedures in some detail. The following section discusses these procedures as they are used in Michigan with comparisons made to the procedures of other regions.

LOAD FORECASTING PROCEDURES

Predicting maximum load for an electrical generating system years into the future is a difficult problem because of the many random factors influencing power use. Some contributing factors, such as population growth or average electrical use per customer, are reasonably predictable because they show a definite historic trend. On the other hand, other equally significant factors, such as economic growth or climatic conditions, show random variations, making it difficult to predict accurately their behavior. Consequently, any procedures for predicting future electrical demand is inevitably dependent on more or less subjective assumptions regarding the future overall trends of the less predictable factors.

In Michigan, the peak load (megawatts) is determined by forecasting electrical sales (megawatt hours) for each of the three major user groups: residential, commercial and industrial. A load factor is defined for each group:

$$\text{Load Factor} = \frac{\text{Average Demand (MWe)}}{\text{Maximum Demand (MWe)}}$$

where   Average Demand   =   annual sales per year, and
       Maximum Demand =   observed demand at time of system peak.
The maximum demand or load is then given as:

$$\frac{\text{Sales x Load Factor}}{8,760}$$

where   8,760 is the number of hours per year. Since each user group has different sales and load factor characteristics, forecasts are made for each group independently.

The method of determining sales projections uses both historical population trends and census data to calculate the average use per customer. Often the user groups are broken down into temperature-sensitive and nontemperature-sensitive users and calculated separately. The temperature-sensitive groups would be predominantly users of air-conditioning and electrical space heating. The commercial and industrial users' average sales predictions are based not only on extrapolation of historical trends but also by use of economic forecasting. Thus electrical sales are related to U.S. economic growth through Gross National Product (GNP) figures. Some utilities sell power in large quantities to a single type of industry and this often has an effect on sales predictions. For example, 35-40% of Detroit Edison's industrial sales go to the auto industry so forecasts are based on that industry's predictions of future sales. Similarly, Consumers Power bases its sales projections on the expectations of sales fo Dow Chemical Company and the GM Truck Division. These expectations are, in turn, obtained from the projections of the respective companies.

The Arab oil embargo of 1973 shows how precarious forecasting methods are. It might have been expected that a scarcity of petroleum products for energy would have led to a greater reliance on coal-fired electrical power and, therefore, to a surge in electrical power demand. Indeed some of this may have occurred; however, the much greater reduction in overall economic activity and lowered industrial output completely overshadowed any such effect. The result was a reduction in electric power usage of some 5 to 7% below the expected value. For example, Toledo Edison, which had experienced a 10-year average growth in kilowatt-hour sales of 8% for the preceding 10 years, had only a 2% growth in 1973-1974.[7]

Now that the country's vulnerability to external energy sources has been amply demonstrated, a new uncertainty arises which makes forecasting more difficult still. Energy conservation is now recognized by industry, government and private citizens to be an important approach to increasing reserves and reducing dependence on energy imports. But how extensive will conservation be and how effective will it be in reducing the demand for primary energy sources? The elasticity of demand, *i.e*, the ratio of demand-to-price can be determined for many products. The elasticity of demand for electricity is difficult to determine but is judged to be very low. Then how shall we judge the elasticity of demand for the kilowatt hour of electricity generated by conservation? Although the cost of a conserved unit of energy is deemed very low, it is very difficult to estimate the degree of conservation which will take place. There are very many intangibles which affect the rate of conservation, such as the cost of conservation (*e.g.,* insulation), the degree of voluntary sacrifice and the impact of laws enforcing conservation. Since this country has not yet had an extended experience with conservation measures, the jury is still out on its effectiveness. Whether or not it will have a pronounced effect on our power demand forecasts remains to be seen.

As shown, many subjective assumptions are required to estimate the future effect and behavior of the less predictable factors influencing electrical load growth. For example, what happens after a significant historical event, such as the oil embargo, is over? Does the growth rate return to normal, increase to make up any growth deficit, or does it become permanently lowered? Although it may be too early to tell what the long-term effects will be, current evidence indicates that the growth rates have returned to almost the same values as before the embargo. The Michigan energy center study team felt it necessary that the growth rate be somewhat less than the historical rates and designed their prototype energy center on this basis. Thus, although the Detroit Edison Company and Consumers Power Company, with the concurrence of the Michigan Public Service Commission, forecast growth rates of from 4.9% to 7.2% for 1974 through 1982, the Michigan study assumed a growth rate of 4.7% to 5.9%.

Electrical generating facilities begun at the present time could not be completed before 1984 in the case of fossil-fueled plants or before 1988 in the case of nuclear plants. This means that an energy center begun in the near future must be designed so that its growth rate is compatible with the load growth for the period after 1982, the year the above forecasts end. Load estimates for the period from 1982 through the year 2000 are even more precarious but are necessarily made for the energy center concept. For the Michigan energy center, the generating capacity and its rate of growth is determined by a simple historical extrapolation for lack of a better predictor. Schedules for plant additions already planned and the schedule of power plants to be retired (usually after 45 years of service) have been obtained from the utility companies themselves. The net capacity needed after taking account of planned additions and retirements yields the capacity assumed to be supplied by the Michigan energy center. Figures 1 and 2 show the respecitve projected load for Detroit Edison and Consumers Power as projected for the energy center. As can be seen from these figures, maintaining a generating capacity sufficient to meet the projected additional demand requires that approximately 1,500 MWe of system capacity must be added per year from 1987 through 1994 with 2,000 MWe per year added from 1994 through the year 2000. If the present 20% reserve is to be maintained, the center must add generating facilities at the rate of 1,500 MWe per year from 1984 to 1991, 2,000 MWe per from 1991 to 1997 and 2,500 MWe per year from 1997 to 2000.

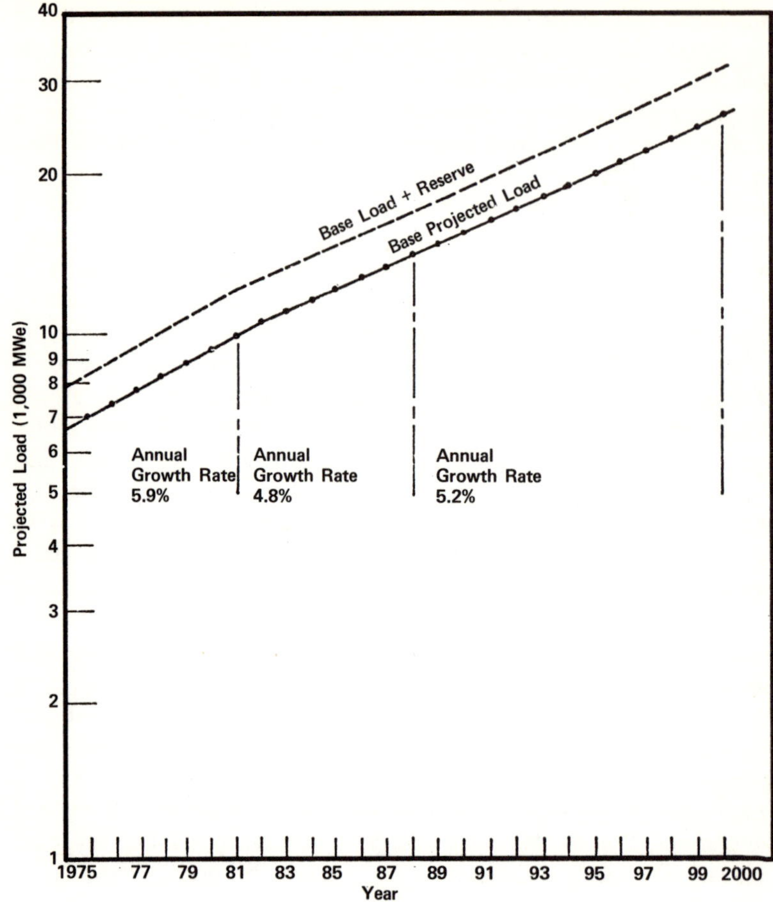

Figure 1. Projected load for Detroit Edison load area.

These load projections are to be considered estimates only because of the uncertainty involved in forecasting. However, it is inherent in the energy center concept that the actual long-term construction schedule depend on updated projections as the construction proceeds.

Figure 3 indicates the combined net growth in electrical load that must be accommodated for Michigan during the 15-yr period from 1985 to 2000. This net growth, including the 20% reserve, is 29,000 MWe. The Michigan study is based on the assumption that a maximum generating capacity of 23,125 MWe will be installed at a single energy center and that new capacity, geographically distributed throughout the system, will

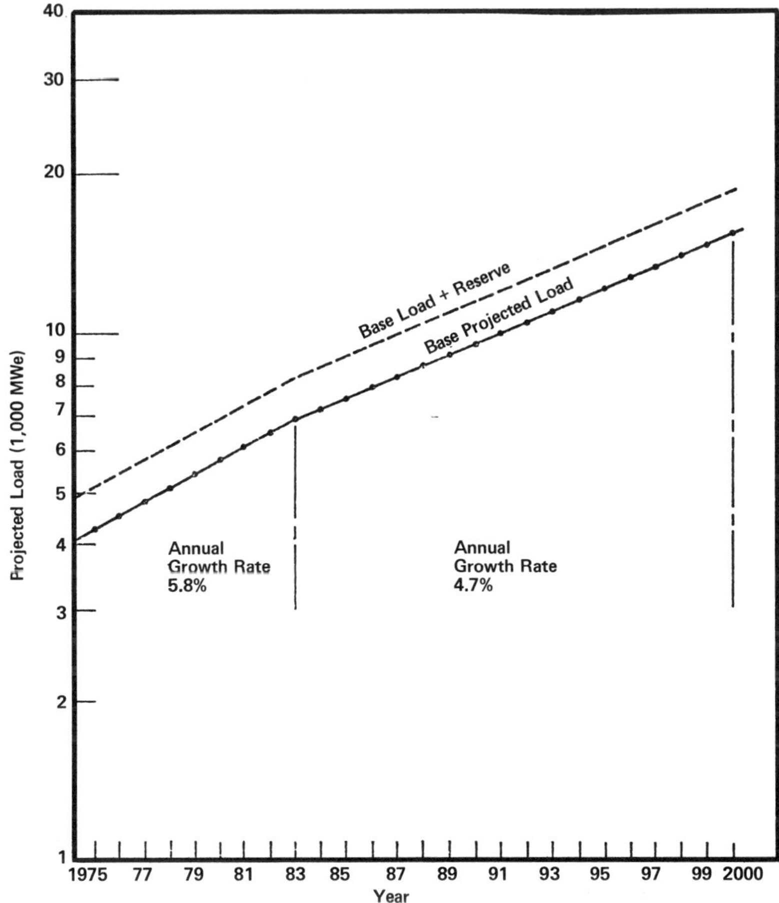

Figure 2. Projected load for Consumers Power load area.

provide the additional 6,000 MWe needed. There are certain advantages
to some geographic distribution of the new capacity; this will be discussed
in a following section.

In the Michigan study it was assumed that fossil-fueled units would be
available on a more rapid time schedule, particularly during the early
period of center construction. Consequently, it was suggested that the
early units be coal-fired. However, the later units were projected to be
nuclear, since these were supposed to provide the lowest-cost power generation. It was further determined that an excess of coal-fired units would

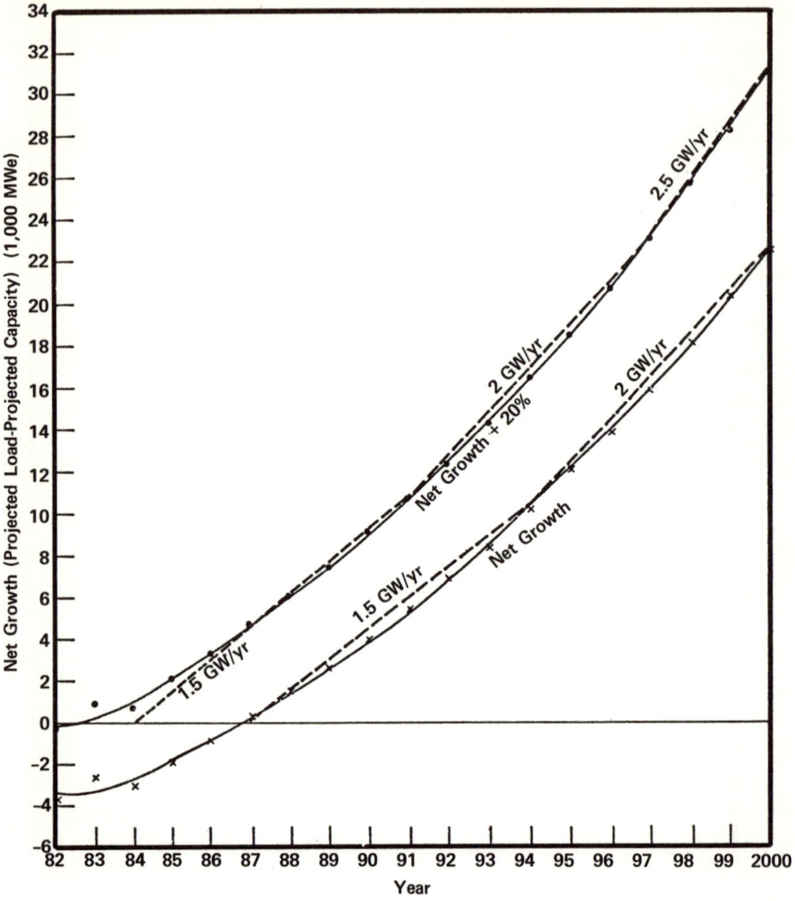

Figure 3.   Center growth required to supply the Detroit Edison and Consumers Power load areas.

probably result in exceeding the allowable discharge of pollutants into the air. Their program thus emphasizes nuclear units during the later years. A division of generating capacity between nuclear and fossil-fueled units also has the advantage of diversifying the generating sources, so that difficulties which might call for partial or total shutdown of one type of unit will be limited to that type rather than to the entire system. This particular problem will be seen, later, to mitigate some of the advantages of any energy center since part of their advantage comes from the construction of many similar units with the concomitant savings inherent in the potential "mass production." At the same time, too many similar units may jeopardize system security.

As a result of these considerations the Michigan study proposed a total generating capacity of 23,125 MWe with one-third fossil-fueled units and two-thirds nuclear units. This mix was selected by the study team primarily for study purposes and was not claimed to be an optimum mix.

THE CAMP GRUBER ENERGY CENTER

Although there has been no specific energy center configuration selected for the Camp Gruber location, there was an attempt to hypothesize the roles an energy center might fill. This led to an approximate size of the facilities needed to meet future demand based on Federal Energy Administration (FEA) forecasts of energy trends, regional needs and the Oklahoma energy-need forecast.

Numerous combinations of demand and supply options have been evaluated by the Federal Energy Administration in their long-term analysis. These options all showed the same basic trends: oil and gas will become increasingly scarce. Shale oil and synthetics from coal, while having a limited impact in the near term, will be needed to make up for the long-term shortfall. In any case, such fuels may determine the price of oil and gas late in the century. Conservation, especially when combined with shifts to electric power, will become imperative in the long run. Increased reliance on electric power will be necessary and this implies dependence on nuclear power or other new sources as well as improved conversion technologies. The FEA further states that the depletion of conventional oil and gas resources dominates the post-1985 period and leads to two fundamental views: synthetic fuels from coal and oil shale will be required after 1985 and their use will continue to grow rapidly. The oil and gas shortfall is so large that major shifts in demand, most probably to electric power, as well as strong conservation measures will be needed.

The assumptions used for the Oklahoma study were based on projections of the FEA and the Southwest Power Pool,[8] and a study by the Oklahoma Energy Advisory Council.[9] These studies led the Camp Gruber team to estimate the size of an energy center for the reservation to be from 15,000 MWe to 30,000 MWe by the year 1990. This is based on the expectation of an 8% annual increase in electrical power demand. This amount of generation would probably be obtained from both nuclear and coal-fired plants. The study pointed out the apparent need for coal gasification and/or liquefaction plants and energy-related industrial plants in the vicinity of the center.

The assumptions used in the Oklahoma study led to an energy-generating facility of about the same magnitude as the other centers examined. However, the projections from the *Energy in Oklahoma* study was somewhat greater than the FEA projections, as well as being greater than the projections for most other centers described in this book. A later study has been completed which shows that the projected loads may be less than assumed for the study and,therefore, the center capacity may be closer to the lower limit by the year 2000.

THE DEMAND AT THE OTHER ENERGY CENTERS

The demand projections used in the other energy center studies are quite similar; only the details and actual numbers used were different. Also, a different weighting was given to the residential, commercial and industrial demand projection in their relation to the importance of these sectors in each region. The results are reasonably similar and led to an assumption of size for the generating facilities of from 10,000-MWe to 40,000-MWe capacity, to service roughly similar-sized regions in the year 2000.

The art of forecasting is exceedingly difficult; it becomes more difficult when the projections are to span longer and longer periods. This difficulty can be aptly demonstrated by comparing the divergent results obtained by different groups and then noting how the divergence grows as the projection span increases (Table II).

Table II. Comparison of Annual Growth Rate Predictions

	Annual Growth Rate Prediction (%)	
Source of Forecast	1975-1980	1980-1985
Energy Industrial Center Study	6.1	4.9
National Gas Survey	7.4	6.9
Electrical World	6.9	6.5
FEA (1974)	N.A.	7.3
U.S. Department of the Interior	7.1	6.7

Note: N.A. = not available.

For the near projection, the rates differ by 1.3 percentage points while for the more distant projection the rates differ by 2.4 points. Although these differences seem relatively small, it must be remembered that these figures are growth rates and behave as compound interest. Thus, in the 25-year span from 1975 to the year 2000, a 5% annual growth in electrical demand will require almost 3.4 times the 1975 generating capacity. On the other hand, a 7% growth rate will require an increase in capacity of over 5.4 times the 1975 capacity, or almost 60% more power plants. In the U.S. at the present time there is a peak generating capcity of about 500,000 MWe. If one concedes that the growth rate will proceed at the historical 7% rate (and there is no evidence yet that it will not) then it will be necessary to add over 2.0 million MWe to our capability by the year 2000! This could be obtained by constructing about 1,400 nuclear plants of the largest size now currently being licensed. Or alternatively, this implies that, on the average, each state in the union would need its own 30,000- to 40,000-MWe energy center. The magnitude of this projected need has led many to doubt that such a growth rate can be sustained.

Even at the lowest projected growth rate, a new generating capacity of nearly one million MWe must be constructed! Again, either 800 to 900 dispersed nuclear or coal-fired plants, or a midsized energy center for every state, is needed. A still lower 3.5% growth rate will require again as much generating capability in the year 2000 as exists now.

Although the application of compound growth rates produces almost unbelievable projections, it also poses another dilemma. This disbelief can lead to inaction which can produce serious, indeed, dangerous energy shortages. If the higher growth rate in demand continues, then the country must duplicate its current generating capability in just ten short years. At the same time it takes ten years from conception to online operation for the typical nuclear plant, and this magnitude of increase is simply not in the current plans. In a way then, it is already true that this country will not, cannot, maintain the historical rate of growth. Whether this will cause serious economic disruption or national danger will depend on new energy sources (solar, geothermal), the results of energy conservation measures and, most probably, changes in lifestyles. It will also depend on further feasibility studies and plans for constructing energy centers.

These projections in electric power demand not only point out the tremendous construction that would have to take place, but also focuses our attention on the very rapid depletion of scarce resources. As far as the power industry is concerned, the scarcity of oil and gas came to the country's attention during the Arab oil embargo. Up to that time the utility industry was experiencing a shift away from coal-fired boilers to oil-and gas-fired. The two main pressures for this tendency was to reduce air pollution by use of the cleaner fuels and the need for more units capable of supplying peak loads. In the first instance, air pollution standards promulgated by the U.S. Environmental Protection Agency (EPA) would require the utility industry either to install costly sulfur-removing stack scrubbers and other pollutant-removing devices, or else to shift to the much cleaner burning oil and, sometimes, gas. The utilities chose the latter approach mainly as a result of nudging from the EPA, the pressure of rising costs and also, perhaps, from the relief of not handling the solid waste disposal products resulting from coal-burning plants. In spite of the fact that coal-fired burners cannot economically be converted to oil or gas, a significant amount of oil- and gas-generated electricity was produced prior to 1974 as many new units brought online were of the oil- and gas-fired design.

SURVEY OF DECLINE OF RESOURCES

Compound growth and the exponential decline in resources has almost become a cliche. It has not been too difficult to convince most people that the traditional sources of energy will be depleted no matter what the actual rate may be. It is clear that these sources will be depleted even with no growth whatsoever. It is a much different matter to predict how long energy resources can be made to last. Estimates of proven and available reserves often differ by orders of magnitudes depending on the assumptions used in making the predictions. Again we are faced with the terribly difficult job of forecasting. For example, in determining the potential energy reserves for fossil fuels the results depend greatly on future costs of exploration and production, the future market price of the fuel, the intensity of geophysical prospecting, the discovery of new sources, the ever-changing rates of consumption, environmental concerns, lack of forward planning, international politics, unforeseen surges in consumption from crises such as war, and many, many other imponderables. Still using the most optimistic assumptions, the picture for the only-too-near term is not very bright.

Natural gas, that most splended fuel, currently supplies one-third the total energy requirements of the U.S. as reported by the American Gas Association. Gas is the cleanest fuel, is transported most easily and is often the most convenient (and for some industries critically necessary). The Federal Power Commission (FPC) has categorically stated that evidence submitted to the commission (in 1971) "confirms beyond any doubt, if indeed there is any remaining doubt, that a serious gas-supply shortage does in fact exist throughout the nation's gas-supply areas." There is a serious shortage of supply, but it is still difficult to determine just how much in-the-ground reserves exist. One reason for this difficulty is that, although the shortage has been apparent for some time, exploration and drilling activities of the petroleum industry declined markedly from the early 1960s on. Wildcat drilling, which is sensitive to market demand, decreased by 40% between 1956 and 1970. And further, more than 200 drilling rigs were removed from the U.S. and transported to more attractive business opportunities in other areas of the world. Figure 4 shows the steady decrease in wildcat wells from 1956 through 1970.[10] In addition the demand for natural gas during the decade of the

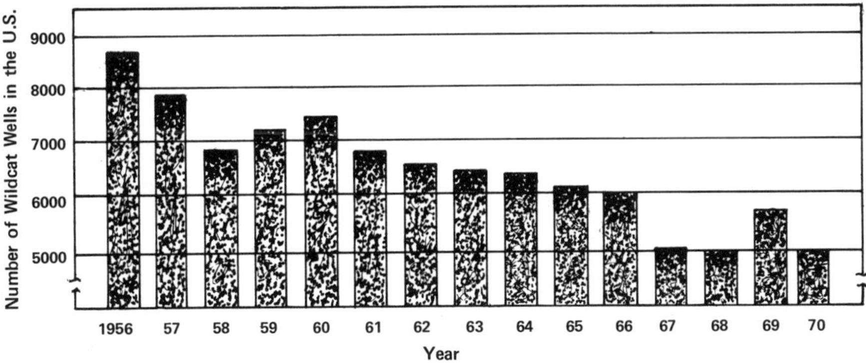

Figure 4. Bar graph showing the decline in the number of wildcat drilling operations.

1960s running from 6% to 8% increased as much as 10% after 1970. Thus the ratio of reserves to production, measured in the number of years of reserves remaining, has decreased from 21-years supply in 1956 to less than 14 years in 1970. This means, simply and frighteningly, that since the beginning of modern industrial society a consumer may have the opportunity to buy a mechanical product whose life expectancy is greater than the expectancy of its primary fuel supply! This may not yet be true of the consumers' home gas clothes dryer but it is most certainly true of the utilities' peaking gas turbines which have a life expectancy or payout time of 20 to 30 years.

Other scenarios or models of future gas production may be slightly more optimistic, but do not substantially alter the eventual outcome. For example, investigators at Consumers Power Company[11] have modeled a situation in which they show how 2,000 trillion ft^3 of supposed natural gas reserves could maintain production at current levels to beyond the year 2010. This slight extension in the life of our natural gas resource would require very extensive new exploration and the concomitant increase in the cost of gas. In the words of the authors of this report this goal is "believed unattainable."

Oil production, which provides about 43% of our current energy requirements has a somewhat greater reserve expectancy. Current expected or estimated ratios of reserve to production range from 20 to 55 years. This estimate is based on the unlikely premise of energy self-sufficiency in the U.S. It has been estimated that the U.S. has approximately 50 billion barrels of remaining proven and prospective reserves and perhaps another 70 billion barrels of undiscovered potential.[12] At current rates of consumption, this country uses something on the order of 15 million barrels per day. The higher figure, then, assumes not only a decreasing rate, but also extraction of all reserves and potential. The lower figure would occur with present consumption rates and the realistic possiblity that not all potential will be realized. Since large-scale plans to use other sources of energy (coal, solar, geothermal, oil shale) have not been produced at this time, and since facilities cannot become operational for 20 years or so, it is clear that the U.S. is rapidly using up its share of the oil and gas, and must in the future rely on other countries to a greater and greater extent.

Coal is our most plentiful source of energy but, unfortunately, it is the most polluting fossil fuel. Estimates of reserves in this country usually range around 500 years of use. Nevertheless, it too can be considered a declining resource from an economic viewpoint; it will be expensive to clean the coal or the air, coal requires prodigious amounts of water for production and use, and reclamation of land after strip mining is costly. For example, predictions of coal prices project a mine mouth cost of almost $11 per ton in 1985 compared to about $4 in 1968. At this rate it may be worth its weight in gold long before the supply expires.[10]

Nuclear power, once thought to be the great hope for U.S. energy self-sufficiency, may develop serious fuel shortage problems as early as 1990. And yet paradoxically, this shortage of nuclear fuel may never materialize unless the myriad of developing problems are solved and increased construction of nuclear plants is begun. The nuclear industry, a dream in the 1960s, has been plagued by technical and regulatory problems (which will be discussed in later chapters).

Solar power and other more exotic sources of energy show some promise, but only after we have solved the current crisis. The present cost and length of development preclude such sources from having much near-term impact on alleviating fuel shortages.

It is clear that an economy seriously weakened by the lack of inexpensive energy would not be an economy which could support the research, development and large-scale construction of systems necessary to exploit solar and other forms of renewable energy. Just as it was necessary to have a vital economy to place a man on the moon, so will it be necessary to have a healthy economy to provide the mechanisms for producing cheap energy beyond the next 25 years.

THE NEED FOR AMPLE SUPPLIES OF BOTH ENERGY AND ENERGY CONSERVATION

An approach to extending our energy supplies of fossil and nuclear fuels is to drastically reduce our rate of consumption through energy conservation by increased use of thermal insulation, more energy-efficient building designs, reclamation of recycling waste materials, more efficient home heating plants, lower speed limits, efficient collocation of electric generation and industry, and use of waste heat. Conservation is so very important, not only because of the decrease in the rate of energy utilization, but also because conserved energy is equivalent to the cheapest energy available. However, there is some question whether energy conservation alone is sufficient to alleviate shortages until new renewable sources become practical. A deep economic decline brought about by the shortage or increased cost of energy could have very serious implications for the future of the newer energy sources. For example, solar power, while renewable and nonpolluting, will require tremendous capital for research and development and new mass-production facilities to bring unit costs down to competing levels. Geothermal sources will require extensive exploration and new expensive generating capability. Fusion power, even if it becomes a reality, will require many billions of dollars of development costs and construction capital. In fact, the one common factor in all the new exotic sources of renewable nonpolluting energy is the capital requirements and the concomitant time required for development. If energy shortages or increased energy costs create a slowdown in the economy in the next few years it is conceivable that the capital might not be available for developing new sources for the next century. If this were to occur, it could cause a downward spiraling economy where the lack of inexpensive energy would create a lack of capital, and the lack of capital would prohibit the development of inexpensive energy sources.

Most experts are optimistic about the long term and feel that the new energy sources will be economically feasible. It is the next 25 years that will be critical. In the remainder of this book we will explore the possibility that the energy center can play an important role in energy conservation. We will describe in detail the siting and construction requirements of a center, the hoped-for advantages and the potential problems,

III

CENTER SPECIFICATIONS AND DETAILED REQUIREMENTS

In this chapter we will discuss the details of the electric generation system for the suggested Michigan energy center. In addition, an estimate of the capital, land, fuel and water requirements will be presented. Whenever interesting contrasts occur, comparisons will be made with the other proposed centers. Rounding out this chapter will be a detailed discussion of the comparison of energy centers with its only real alternative in current use, dispersed siting. This is a most important topic because a comparison favorable to energy centers becomes their reason for being. However, the favorable advantage that is characteristic of the energy center is a precarious one and can be lost through careless planning and execution.

MICHIGAN ELECTRIC POWER GENERATION FACILITIES

The Michigan energy center study used for their purposes the characteristics of the nuclear- and fossil-fueled units discussed in the study conducted by the General Electric Company.[13] The turbine configuration selected has a maximum rating of 885 MWe each. The turbine design consists of a 4-flow 3,600-rpm unit with a 33-in. last-stage bucket. The station heat rate for the unit at 3.5 in. Hg back pressure is 9,093 Btu at rated power. The maximum rating of 885 MWe is based on inlet steam properties at 3,500 psia and 1,000°F, with one reheat to 1,000°F. The study design was based on generating systems currently available as state-of-the-art. Estimated process steam requirements totaling about 16 million pounds per hour would be provided by extraction of steam at the cold reheat point of three of the fossil-fueled steam turbine units. This full steam capacity would reduce the electric output of the three (out of eight proposed) units from 885 MWe to 600 MWe each.

Three types of nuclear power units were examined in the General Electric study: a conventionally fueled mini-reactor, a plutonium-burning fast breeder reactor, and a large conventionally fueled reactor. For the purposes of the Michigan study, the last alternative was used to describe the equipment complement. The Michigan study pointed out that this did not imply a firm recommendation for center design, since such a recommendation would require a much more detailed analysis than they were able to provide. Instead, the selection was intended for illustrative purposes only.

Each of the nuclear-fueled steam turbine units selected are rated at 1,300 MWe. This maximum unit rating was, at the time of the study, limited by the thermal discharge ceiling established by the then Atomic Energy Commission. The turbine selected was a 6-flow 43-in. last-stage bucket, 1,800-rpm unit with an inlet steam pressure of 975 psia.

CONSTRUCTION SCHEDULES

A construction cycle of 48 months following start of construction on the fossil-fueled units was projected. For the nuclear units, a construction cycle of at least 50 months following start of construction was hypothesized (it may well be considerably longer at the present time[14]). Individual units of each type will start up at 12-month intervals. Table III shows the schedule for completing individual generating units by 5-year intervals. The Michigan study team intended this construction schedule to approximate the load capacity curve shown in Figure 3 in Chapter II. Based on the 50-month construction schedule and the requirements of working on multiple units, the average work force at the nuclear portion of the energy center was predicted to stabilize at around 3,800 manual construction workers. For the fossil-fueled center, the average work force required throughout construction was put at between 2,000 and 2,500 persons. The operating personnel at full-scale operation of the center was expected to reach 1,600 persons.

Table III. Schedule of Start of Construction of Generating Facilities

Time Period	Fossil-Fueled		Nuclear	
	600 MWe	885 MWe	1,300 MWe	Total
1985-1990	2	2	3	6,870 MWe
1990-1995	1	2	4	7,570 MWe
1995-2000	–	1	6	8,685 MWe
Total MWe	1,800	4,425	16,900	23,125

SYSTEM DESIGN CONSIDERATIONS—
THE MICHIGAN CENTER

The use of identical units in the complement of generating facilities at an energy center has a number of advantages which would contribute to the economy of the installation as compared to dispersed sites. These economies result from such factors as parts interchangeability; common manufacturing and maintenance facilities; reduced labor forces required for construction, operation and maintenance; and easier operator training.

Although the use of a single-design generating unit replicated at the center has been touted as capable of achieving appreciable capital savings, there are at the same time, some disadvantages. First, replicating a design over a 15-year period limits the use of probable technological improvements occurring during the construction period. Second, it is possible that if a serious deficiency was discovered, the Nuclear Regulatory Commission would require a shutdown of all similar units until the deficiency could be corrected. Such an eventuality could have a crippling effect on power availability for the region served during the period of difficulty. It should be noted that the vulnerability to a serious shutdown is not unique to the energy center concept and can occur even in dispersed facilities. However, if design improvements could be made on succeeding units as deficiencies are found, it is possible that the savings inherent in unit replication, capital savings and system reliability could all be maintained. Such a possibility would require a very detailed engineering analysis of system design and operating features.

In the Michigan energy center concept it was proposed to locate nearly all new generating capacity at the one site. However, system reliability, which requires proper voltage regulation and proper distribution of the reactive load, dictates that some generating capacity, geographically distributed throughout the system, must be maintained in operation at all times, even though most of the newer units would be located at the energy center.

In any large-scale electrical power network, it is necessary to support voltages which might reach unacceptably low levels during heavy loading conditions and to suppress voltages during light loading periods. This support is provided by reactive power and is normally supplied by capacitors, inductive reactors and synchronous machines, as well as by building reactive capability into the generators themselves. This total reactive support cannot be effectively delivered to load centers from remote generating points but must be near the point of need. Since reactive support must be distributed throughout the system, some base-load generation should also be distributed to maintain system security and reliability.

These considerations dictate that some base load be scattered throughout the system (possibly by means of the generating units already in use) while some special peaking units and a portion of the 20% reserve capability be located at the center. Fuel for the special nuclear units at the center might be supplied by a coal gasification facility, which will be discussed later in this book. Another alternative is to split the growth in capacity between two or more widely separated energy centers. This alternative will be more thoroughly investigated in a separate chapter.

The preceding discussion illustrates some of the system considerations which would make the detailed design of a center difficult and complex. The Michigan study team, however, felt that the simplified model described in their report would, nevertheless, provide an approximation for a feasibility determination of energy centers. Although these system complexities may cause some reduction in overall efficiency and economy, it was believed that the advantages of an energy center can be achieved.

CAPITAL INVESTMENT

This is a very difficult area in which to make estimates; a section is devoted to the problems of capital investment in a following chapter. The problems of capital—how much is needed and where it will come from—is not restricted to energy centers but must be faced squarely in considering any energy alternatives. For energy centers it may be a more critical problem because the inability to account for cost escalations may wipe out much of the expected savings in capital requirements.

For the Michigan center, the total cost of the first 883-MWe unit of the coal-fired system was estimated at $412.4 million in January 1975 dollars. This estimate did not include escalation of completion cost nor interest during construction. It was further estimated that the equivalent cost of the 1,300-MWe nuclear unit would be about $520 million depending on the design selected.

The total capital costs for the complete center are shown in Table IV. Although these costs were computed with the assumption of no escalation or inflation, they more clearly represent costs in current dollars.

LAND REQUIREMENTS

Table V gives a summarized breakdown of the land requirements for just the electric generation facilities of the energy center. A total of 9,510 acres has been estimated. The waste disposal and storage area for the energy center was sized to accommodate the entire 40-year life of the the center. This assumption considered that the present level of waste

Table IV. Estimated Costs of Michigan Energy Center
(Billions of 1975 Dollars)

	Coal-Fired Units	Nuclear Units	Total
Capacity (MWe)	6,225	16,900	23,125
Capital Costs			
Power Plant	2.67	6.06	8.73
Transmission	0.26	0.62	0.88
Land	0.05	0.09	0.14
Gas Turbines	0.01	0.03	0.04
Total	2.99	6.80	9.79
Interest during Construction	0.20	0.50	0.70
Total Present Value of			
Revenue Requirements	4.39	7.56	11.95

Table V. Estimated Land Requirements for the Michigan Energy Center

	Fossil-Fueled Section	Nuclear Section	Total
Number of Units	8	14	22
Total Capacity (MWe)	6,225	16,900	23,125
Land Use (ac)			
Generating	430	1,100	1,530
Cooling Towers and Makeup Pond	720	2,900	3,620
Fuel Storage	160	–	160
Waste Disposal	3,600	–	3,600
Construction Area	300	300	600
Total	5,210	4,300	9,510

accumulation from coal-fired plants would be reduced by about one-third based on an expected improved scrubber technology, recycling of scrubbed wastes and utilization of other waste materials. For example, fly ash can be used as a fertilizer adjunct, a filtering agent or as a building material aggregate.

COMPARISON OF ENERGY CENTER
WITH DISPERSED SITING

The overall advantages of energy centers compared with the dispersed approach to generating the same power has been investigated in the definitive study by the General Electric Company[16] completed in May 1975:

> The overall objective of the study was to examine and compare the techni-
> cal, economic, environmental and institutional issues related to the energy
> park concept, and to identify the obstacles, benefits and penalties that would
> result if the concept were adopted. . . . The time frame of the study assumes
> initial generating unit startup in 1985, and completion of construction about
> 20 years later.
>
> The reference energy parks consist of either 20 nuclear (light-water reactor)
> units each generating 1,300 megawatts-electrical for a total of 26,000 MWe
> or 24 fossil (coal) units, eight each of 885 MWe, 1,075 MWe and 1,320 MWe
> for a total of 26,240 MWe. The progressively larger fossil units reflect ex-
> pected growth in fossil-generating facilities during the period covered by the
> study. The selected total capacities are larger by a factor of four or five than
> several evolving multi-unit generating plants and represent sizes that should
> reveal the potential issues, problems and advantages associated with the energy
> park concept.
>
> The dispersed sites used in the comparative assessments each have roughly
> one-tenth the capacity of a park (2,600 MWe in the nuclear case, 1,770-
> 2,640 MWe in the coal-fired cases).
>
> The study was undertaken on a nonspecific site basis. Energy park and dis-
> persed site locations. . . . are intended to be illustrative in nature.
>
> The study was conducted on the basis of generally current or readily fore-
> casted technology.

In the General Electric (GE) study, the nuclear units (all 1,300 MWe),
reflected the maximum reactor size permitted by regulatory agencies.
Successive units were assumed to become operational at one-year intervals.
Natural draft evaporative cooling towers were used for all generating units
for the comparative evaluations.

Land requirements for the energy park generation sites were estimated
to range from 6,000 to 42,000 acres, depending on the cooling mode and
waste disposal methods. Although land requirements for transmission are
greater due to the greater distances involved, future pressures would also
force dispersed generating facilities to more remote locations. Such a
trend would diminish the large land and cost penalty associated with long-
distance power transmission from energy centers.

The GE task force estimated the magnitude of savings in labor, material, interest and front-end costs resulting from modular fabrication and
construction techniques, quantity purchasing of materials and equipment,
a more stable labor force, standardization of facility design, shorter construction times and integrated support and maintenance procedures. In
many instances these savings were substantial.

The GE study determined no clear-cut technical or institutional obstacles
to the energy park concept. Although many problems remain unsolved,
their analysis indicated that energy centers are technically feasible with
probable economic and other benefits. Overall, the GE task force comparison favored nuclear parks over dispersed sites. They did warn that

flexibility in design and installation of the facilities was necessary in order to adapt to newly emerging problems. Although such flexibility runs counter to the achievements of standardization, it was believed practical to include flexibility in schedule, equipment and system details without nullifying the overall economic advantages.

A most important finding of the GE study was that the use of an onsite factory and modular production-line-type construction indicated a potential 25% capital cost savings for the reference park compared to dispersed sites. In fact, capital savings were anticipated for much smaller-sized parks also, at least down to a 10,000-MWe capacity. These savings result from the standardization of units. Departure from this standardization to provide flexibility beyond minor changes or small corrections of design errors may well jeopardize portions of the expected economies. An expected increase in productivity by 1.5 to 4 times would accrue from shifting labor from the field construction site to an onsite modular factory, combined with benefits from repetitive or "learning-curve" work sequences.

When considering the total revenue requirements for the 26,000-MWe nuclear park, there was an estimate of a 10% saving over the 10 two-unit dispersed sites with the same total equivalent capacity. For the coal-fired units, this savings was about 3%.

The study team from GE concluded that integration of the nuclear-fuel cycle in nuclear parks would provide special materials safeguards and security in the offsite transportation of these materials. If plutonium-burning light-water reactors are desired, there appears to be an advantage in having all plutonium handling and recycling confined within the park.

In comparing energy parks with dispersed siting the GE energy team noted the advantage of collocating industrial facilities near the park. This advantage stems mainly from the continuing assurance of an energy supply for the industries. As far as the reliability and security of the electric power supply was concerned, the task force concluded that there was little difference between energy parks and dispersed siting.

In the area of environmental considerations energy parks begin to lose their advantage. Environmental consideration of water supply, air pollution control (in the case of coal), thermal effluent control and climate effects tend to favor dispersed sites, since any negative effects are more noticeable when large amounts of generating capacity are concentrated at one location. The question here is: are effects on the environment, which are concentrated and therefore more noticeable, more damaging than the same effects dispersed over a wider area? It is clear that 25 cases of emphysema traced to the polluting effluents of an energy center would cause greater outrage than 25 similar cases scattered through the country.

This may just be the nature of things but it does raise the moral issue that dispersed catastrophy may be more acceptable. It does seem to be in "the nature of things," and it is inherent in the energy center concept, that concentrated problems are more keenly focused and therefore unacceptable.

The GE study determined that air pollution requirements, with current technology, present constraints on the size (capacity) of a coal-fired park. Depending on implementation of the "no significant deterioration" regulations under the Clean Air Act, there will be a maximum allowable concentration of large coal-fired generating units. In addition it appears that logistical problems of supplying sufficient quantities of coal at one location appear to limit the size of a coal-fired park.

Another problem area in the implementation of an energy center concerns the so-called "institutional issues." This category includes the organizational, regulatory and legal problems arising from the construction and operation of an energy center. A case could be made that the difficulties lie in the deficiencies of the political systems rather than in the concept of the energy center; this may well be true. Nevertheless, the present state of local, state and federal governmental organization is such that substantial structural changes must be instituted in these political bodies. This is in addition to the need for strong support and initiatives required of the major governing organizations. It is not clear at this time whether major leadership must come from the federal, state or local level. But it is clear that a strong impetus must spring from some, or maybe all, levels in order to bring about the many advantages for energy capital savings inherent in an energy center.

CAMP GRUBER CENTER SPECIFICATIONS

Although the Camp Gruber study team did not fix the size of the proposed Oklahoma energy center, many characteristics were assumed to be similar. For example, their center size would be at least 10,000 MWe and would consist of some mix of coal- and nuclear-fired units; the total quantity of energy handled in the energy center would be at least 30,000 MWe. On this basis the Camp Gruber study team defined a typical work force in order to estimate the economic impacts of such a labor force on the surrounding communities of Muskogee, Fort Gibson and Braggs. They considered several construction schedules for the energy center including: (1) construction of energy utilization facilities, plus numerous large electricy generating plants, so that one 1,250-MWe-capacity plant would start each year for several years, and (2) a large nuclear fuel enrichment plant and associated 1,250-MWe power plants. For Camp

Gruber, it was considered sufficient to examine the availability of a construction force since an operating force is usually much smaller.

In the Camp Gruber analysis, it was assumed that actual construction of the center would start in 1980. During the three years the feasibility of the center would be demonstrated, the decision to build would be made, and the institutional problems of establishing responsibility and procedures for building and administering the center would be resolved. Selection of the first facility and preconstruction activities would result in physical construction beginning in 1982. The peak construction force of some 6,000 workers would then be expected about three years later in 1985.

The Camp Gruber study team observed that since the duration of the peak force determines the quantity of tax and business sale income, which helps repay the cost of facilities needed to accommodate the construction workers, the worst situation would be a short-duration peak force with rapid reduction of the construction force to zero thereafter. The best situation would be a long-duration peak force with a longer time period to pay for the facilities. A more stable construction employment or longer construction period would mean less social impact.

IV

TRANSMISSION CONSIDERATIONS
FOR ENERGY CENTERS

In this chapter we will discuss the special problems of delivering electric power from an energy park to the load areas, especially in those aspects which differ from dispersed generating facilities. These aspects include the planning and operation of a transmission system, economic comparisons, load flow, stability and transmission security. Transmission facilities tie the power system together so that the generating capacity at the energy center may economically supply the system load, which is usually scattered over a broad area.

Transmission line voltages for transmitting bulk power in the U.S. are nominally 345, 500 and 765 kV, depending on the size and distances of the load. In the normal three-phase system, these are the voltages that exist between the conductors. The power delivered to a load is the product of the (in-phase) voltage and the current. Electric power transmission is not 100% efficient and the power loss in a transmission line is the squared current times the lines resistance, the I^2R loss. Since a higher-voltage transmission line can deliver the same power with less line current, the I^2R losses are less and the line can be more efficient. However, delivering the same power with higher voltages and lower currents requires higher towers and larger insulators; the higher voltage lines cost more to construct. There is, then, a trade-off between line losses and line costs that is dependent on the size of the load to be delivered.

Delivered power consists of two components referred to as real power (the product of the in-phase voltage and current) and the reactive power (the product of the out-of-phase voltage and current). Real power supplies useful work, heat, light and mechanical movement. Reactive power is a component which is transferred back and forth between the inductive and capacitive elements of the system. The inductive and capacitive elements

37

of any network are essentially energy storage elements which transfer energy between each other though no actual work is done. A typical inductive load is an electric motor, while capacitive reactance is often associated with long underground cables. The transfer of energy between such network elements will increase the current flow and thus require a larger transmission system than for real power. The reactive power also increases the line voltages during switching operations and thus needs additional insulation. It is clearly important to reduce reactive power as much as possible. This is accomplished by installing capacitors near inductive loads, such as electric motors, and inductors near capacitive loads, such as underground cables.

The load-carrying capacity of an alternative current (AC) transmission line is usually conveniently defined in terms of the surge impedance loading given by:

$$P_{SIL} = V^2/Z_0$$

where V is the line voltage and Z_0 is the surge impedance. Surge impedance is defined as the ratio of voltage-to-current in a transmission line such that the capacitive reactance power produced by the line is just compensated for by the inductive reactance power the line consumes. Under these circumstances the surge impedance is given by:

$$Z_0 = \sqrt{L/C}$$

where L is the inductance and C is the capacitance of the line per unit length. At surge impedance loading, the voltage everywhere on the line is constant and there is no net reactive power present or generated. If an uncompensated line is loaded above surge impedance loading there is a specific line length beyond which the reactive power that must be supplied to the line rapidly increases. The maximum line length for a given loading P is given by:

$$L = \frac{c}{\omega} \sin^{-1}(P_{SIL}/P)$$

where c is the speed of light, ω the angular frequency of the current (about 377 for the standard 60 Hz or cycle per second system), and P_{SIL} is the surge impedance loading of the line. This consideration places a maximum length on any transmission line with a specific loading. For example, the maximum length of a 60-Hz line operated at 1.5 times surge impedance loading is 360 miles. The relation of line load with respect of line length is given in Figure 5.

The most critical limit on line length or loading is imposed by the conditions for stability of synchronous operation. If an AC system is loaded at or above the "steady-state stability limit" the generators go out of synchronism with the load and can accelerate and generate large

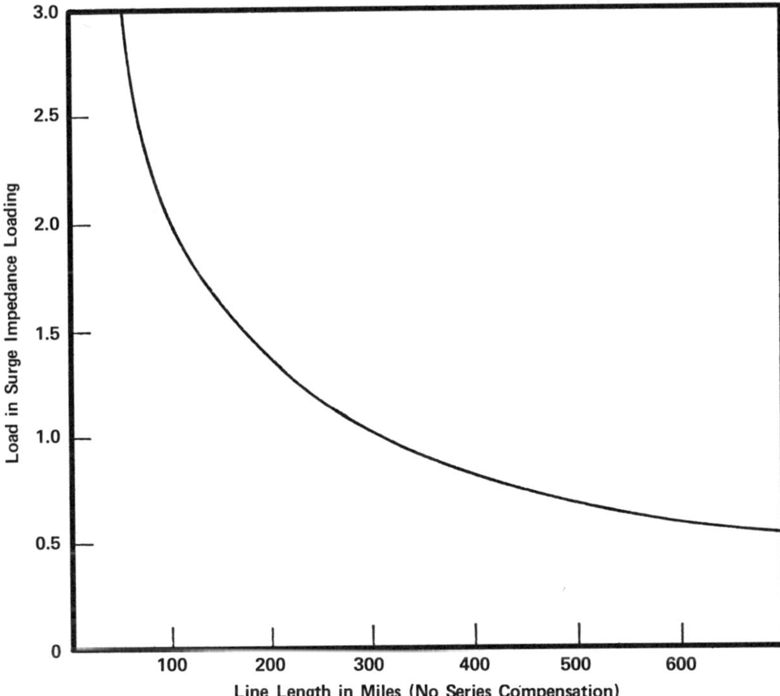

Figure 5. Typical line loading as a function of line length taking into account stability criteria.

voltage fluctuations in the system. The "steady-state stability limit" is given by:

$$P_{MAX} = \frac{V_1 V_2}{X_L}$$

where V_1 and V_2 are the voltages at the sending and receiving ends of the network and X_L is the total inductive reactance of the system. Even when a line is loaded below this "limit" it can become unstable as a result of a large transient or perturbation on the system, such as a short circuit or lightning hit. A margin of safety is usually maintained by keeping line loading less than or near one-half this limit.

Another important factor governing the number of transmission lines is the consideration of reliability. Reliability, as the term implies, is the ability of a system to supply power reliably, even when unusual conditions (outages) occur somewhere in the system. The general rule used by most power companies is that the system should be able to suffer the loss of any major line, generator or bus section without overloading remaining lines or suffering instabilities. This means that for every circuit loaded at or near the transient stability limit, a second circuit should be available which can handle the load should an outage occur.

Planning a transmission network for power delivery, whether by dispersed sites or from an energy park, requires that one determine (1) how much power is to be transmitted, (2) how far it is to be transmitted, and (3) how reliability is to be maintained. The most distinctive difference between dispersed siting and energy centers is that energy centers almost always will require longer transmission distances. Stability considerations become increasingly important with distance. Either a relaxation of system stability must be accepted or reduced ratings and costly compensation will accompany power delivery from energy centers remote from the serviced load.

Another serious fault of a power transmission network is the short circuit. This is an unexpected surge of current caused, for example, by a downed line. To protect against short circuits, the system is equipped with circuit breakers which disconnect the generators from the fault quickly enough to prevent damage to the system. At the current state-of-the-art, it is possible to quickly shut off currents of a maximum 50,000 A. Any short currents larger than this would create damage to the protective equipment or the generators. Thus the generating capacity of an energy center is limited by the amount of short-circuit current that the breakers can interrupt when a fault occurs. For the proposed 765-kV transmission lines this does not seem to pose any special problem for the energy center. In particular, if the center is located somewhere near the geographic center of the load, then there will be more than one bus supplying the load and each will be able to handle any eventual short circuit.

An electric power transmission system must be designed so that all generators remain in synchronous operation under all operating conditions, steady-state and transient. Transients include short circuits as well as sudden reductions in all or part of the load. In order to maintain the 60-Hz alternating current in the transmission system, all generating units must be synchronized, that is, the speed of rotation of these machines must be carefully controlled. If there is a tendency for a machine to slow its rotation, more steam pressure is required to bring the rotation back in sync with the other generators. If the change in load occurs slowly, the turbine governors can react quickly enough to restore the generator speed to normal. In the case of a transient or very rapid change in load, it may be difficult for the system to react rapidly enough to keep the generators in synchronism. If, for example, a transient increase in load affects a generator in the system, it may cause it to slow down and generate current at 58.5-Hz frequency instead of the proper 60 Hz. In such an event it is possible that very large currents are exchanged between machines rather than being delivered to the load. These exchange currents can be large enough to trip protective mechanisms which isolate the generators from the system. But this sudden loss of generator capacity is itself a transient and causes further demands on the already overloaded generators remaining connected to the system. The ability of a power system to react quickly to load changes and maintain all generators in synchronous operation is called system stability.

The ability of a system to adjust to load changes, its stability, depends on the speed of system reaction and control. This speed of reaction to a loss of synchronism is related to the rotational inertias of the connected machines, the rate of change of the load, and the rapidity of the responses of the generator exciters, the turbine governors and the control system. System stability also depends greatly on the reactance of the transmission line which in turn is related to the distance between source and load. Stability considerations thus partly determine the maximum transmission distances and may, therefore, be a more severe constraint for energy centers than dispersed siting.

COMPARISON OF TRANSMISSION FACTORS BETWEEN ENERGY CENTERS AND DISPERSED SITES

A comparison between energy centers and dispersed siting depends on cost differences, transmission line losses, stability, reliability and other factors. Although it is exceedingly difficult to prove a clear advantage for one approach with regard to transmission considerations alone, it will be of interest to detail this comparison.

One comparison between energy centers and dispersed siting is the choice of available technology. Recent advances in AC-transmission technology have increased the voltage of the lines with an attendant increase in line capacity for the same percentage power losses. 765-kV capacity lines are now in common use. Table VI provides some comparisons among 345-kV, 500-kV and 765-kV lines. The costs given are estimated 1975 costs and may not be representative of today's costs. Nevertheless, it is expected that a comparison based on cost ratios will remain approximately the same.

Although the totals in Table VI do not include line compensation costs and right-of-way acquisition costs, they do show how the higher voltage lines are more expensive on a per mile basis. However, what is important is the amount of power than can be transmitted by the line and the

Table VI. Capacities and Costs of Overhead Transmission Circuits

	345 kV	500 kV	765 kV
Voltage	345 kV	500 kV	765 kV
Capacity for 250-mi Length	625 MW	1,500 MW	3,300 MW
Land: Corridor Width	55 m	64 m	75 m
(ac/mi)	18	21	25
Line Cost (million $/mi, est.)	0.120	0.230	0.419
Terminal Cost (million $/line, est., excluding transformer)	8.83	12.95	26.00
Cost Increase for Energy Center Over Dispersed Sites (million $/line)	–	2.0	20.65
Transformer Costs (million $)	5.63	13.5	29.70
Total for 250-mi Line (million $)	44.46	85.95	181.10

transmission costs of this power. Since the higher-voltage lines have a much greater power capacity, this offsets the total installed line costs. Thus the installed costs in $/kW are around $75/kW for a 765-kV circuit compared to a 345-kV line costing over $120/kW (in 1976 dollars). This last comparison determines the cost of delivered electricity and shows the advantage of higher-voltage transmission lines whenever they can be operated near capacity. An energy center with its requirements to deliver large amounts of power over long distances can effectively make use of the economic advantage of the high-voltage circuits.

Technology is becoming available to increase the voltage of the transmission lines even further, say up to 1,100 kV, with the concomitant higher power capacity (about twice that for a 765-kV line) and a decrease in costs per kilowatt. Would these high-voltage lines (referred to as UHV lines) serve an energy center better? Probably not, since most plans for energy centers call for a rather slow increase in generating capacity to meet only the growing demand and the 765-kV lines seem to be a better match to the proposed incremental growth, as well as provide system reliability through redundancy. The tremendous load-carrying capacity of the 1,100- to 1,200-kV lines is over 5,500 MW. Even if four 1,300-MWe generating units were installed simultaneously, it would only partially load one line though two would have to be constructed for system reliability. This type of instant growth is not envisioned for energy center development.

HIGH-VOLTAGE DIRECT CURRENT TRANSMISSION LINES

Alternating current transmission of electric power is currently predominant because of the ability to increase voltages for economical long-distance transmission and the ability to decrease voltages at the load to usable levels. This step-up and step-down capability is accomplished through transformers. Direct current (DC) voltages cannot be stepped up or down by these transformers. Nevertheless, there is new technology which can be used to convert voltage levels of direct current so as to exploit special advantages of DC power transmission. The advent of High-Voltage Direct Current (HVDC) solid-state terminal equipment has provided designers with a practical alternative to EHV and UHV transmission. Several installations such as the Eel River tie, and installations in central North Dakota, British Columbia and Manitoba, are being used to evaluate the practicality and economics of HVDC.

HVDC transmission of power becomes competitive when it is necessary to transmit large quantities of power long distances. The cross-over point currently appears to be greater than about 400 miles for 2,000 MWe of power. Since this is one of the special transmission requirements of an energy center, it would appear that HVDC may have an economic advantage. Besides the cost factor, HVDC has other advantages which may be especially useful for energy centers. The cost advantages come from the increased line capacity and, in some applications, lower cost per MW-mile using DC. Also fewer lines (2 as opposed to 3) are needed for DC than AC. Right-of-way costs and underground cable costs are also lower for DC. On the other hand, DC terminal costs are considerably more expensive than AC. This is the main reason that AC is still favored for shorter transmission distances. Environmentally and esthetically, DC appears to be the more attractive alternative because (1) it requires smaller lines and rights-of-ways, (2) the terminals, being solid-state, lend themselves to miniaturization, and (3) visibility would be lower without the bulky transformers.

System reliability and stability, particularly important for energy centers, may be greatly enhanced by the use of HVDC. In the case of reliability, HVDC is superior because it is less susceptible to conductor outages (smaller lines and fewer conductors). It is possible to operate DC with one conductor and an earth return whereas a line-to-ground fault on AC will cause a total line outage using today's systems. In addition, DC faults are not catastrophic in the sense that large short-circuit currents do not develop because of the controlled power flow nature of DC and line faults have the potential of being sensed, cleared and restored within 0.4 seconds. DC has lower dielectric stresses than AC and, therefore, is less affected by insulation problems.

An increase in transmission network stability may be one of the biggest advantages of HVDC. In an earlier section, it was pointed out that AC generators required synchronous operation to maintain frequency, phase relations and to prevent the flow of large currents during transients. Long AC transmission lines have inductive and capacitive reactances which require the generators to supply reactive (not "real") power to maintain proper phase relations. When transients (sudden load changes, line outages or generator losses) occur, the electrical length of the transmission line changes, possibly causing large power surges which can alter generator speed, causing a loss of synchronism and leading to the necessity of tripping overload relays, which may cause further transients. Such an instability can propagate throughout the network quickly, causing power loss of wide areas. In HVDC the generation equipment is not subjected to subsynchronous resonance problems associated with AC lines. DC lines, then, do not have the overcurrent or overvoltage surge problems characteristic of AC due to switching and line failures. In a large tranmission network, system stability conditions are improved with the addition of some controlled DC lines. These considerations allow the reduction of transmission margins and reserve capacity.

Although there are some disadvantages to HVDC, the ease with which large amounts of power can be consolidated or pooled suggests that it would be particularly appropriate to the energy center concept. Since energy centers have additional requirements of reliability, stability and the transmission of large amounts of power over long distances as compared to dispersed siting systems, it appears that HVDC transmission systems and their special advantages in these areas lend themselves to the concept. Any advantages centers have over dispersed siting may result from the appropriate design of a HVDC transmission network.

TRANSMISSION DESIGN CONSIDERATIONS FOR ENERGY CENTERS

Reliability of transmission is a major consideration for the large energy center concept. It is necessary to reduce the probability of simultaneous multiple-unit outages, which is accomplished by special switching arrangements for units, buses and lines, the ability to isolate units within the complex, and by connecting multiple transmission lines to each generator bus. When several generators are connected to one bus, the loss of the entire bus must be considered a serious potential since the interconnected generators and transmission-line loads can produce large current surges or transients. For conventional generators of about 1,200-MWe capacity, a maximum of four units should be connected to a single bus. At least two transmission lines are required from each generation bus to maintain reliability due to outages. At an energy center, which requires long lines, it may be necessary to connect more than two lines to each bus to maintain system stability. It is a dictum of the electric power industry that complete system service should be maintained through the loss of the largest generating unit in the system. This requires that a switching arrangement be designed such that, for each bus, the loss of one generator will not produce the loss of others. This can be accomplished by the use of double-breaker schemes for machine isolation. These requirements for system reliability suggest a possible configuration of generation and transmission as shown in Figure 6.

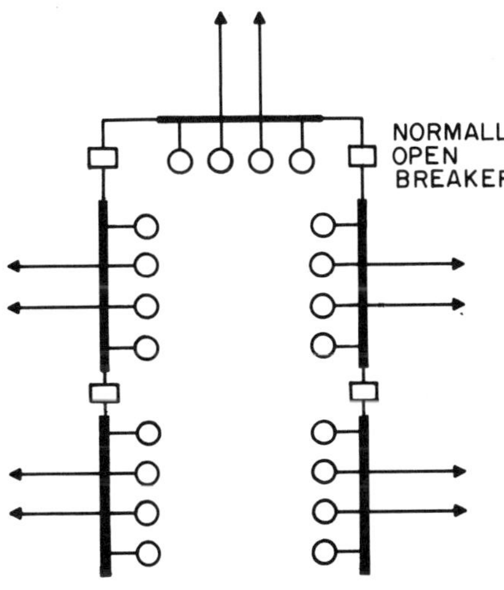

Figure 6. Illustrative electrical arrangement of an energy park.

The electrical arrangement of an energy center shown in Figure 6 illustrates several principals. Here we have five groups of four 1,300-MWe generating units connected to a single bus feeding two 765-kV transmission lines. Each line can handle about 5,000 MWe or nearly the total capacity of all four generators. This provides system reliability in case one line is out. It will also be noted that the buses are connected through normally open switches so that power can be quickly diverted to other buses in the event that one or more generators are out or more line capacity is needed. This arrangement provides the so-called double outage protection, interpreted to mean one circuit suddenly tripped while a second circuit is down on maintenance.[13] To maintain system reliability for a dispersed siting method of power generation, it would be necessary to construct three 765-kV lines for a 5,200-MWe bus rather than two. However, this is not an overwhelming disadvantage for energy centers since transmission costs of power are only a function of distance-to-load, not the proximity to other buses.

As mentioned previously, the system is designed to maintain service if there is a loss of the largest generating unit in the system. Similarly, each generating plant should be connected so that the sudden loss of any single line does not interrupt or limit the output of the center. What if both circuits connected to a single bus are interrupted suddenly while carrying full power output? The system would lose up to 5,200 MWe. In a 1975 system this would be a serious outage (Michigan's total 1975 capacity was only about twice this). However, a center capacity of 5,200 MWe would not be completed by 1990 in most energy center plans and, by that time, 5,200 MWe would be only 25% of the total system capacity. The deficit would be made up from the 20% reserve normally maintained and by purchasing power from neighboring power pools. This example does suggest that whatever the growth rate in the capacity of an energy center, it should never be much more than 25% of the total system capacity.

SUMMARY OF COMPARISON OF TRANSMISSION CONSIDERATIONS BETWEEN CENTERS AND DISPERSED SITES

1. Current transmission system technology is adequate and available to handle power delivery from energy centers of 20,000 to 30,000 MWe.

2. The 765-kV line matches growth plans of a 26,000-MWe plant and maintains security better than UHV.

3. Overhead transmission costs for an energy center would be about twice the cost for dispersed siting.

4. Overall system reliability can be as high with energy centers as with dispersed siting.

5. System stability can be made comparable to dispersed siting or even better if HVDC is used.

6. HVDC may be preferred for transmission distances over 400 miles.

SELECTION OF TRANSMISSION NETWORK FOR MICHIGAN ENERGY CENTER

Figure 7 shows the state of Michigan divided into load areas as defined by Consumers Power and Detroit Edison along with the percentage of the total Michigan load each area consumes. If it is assumed that these

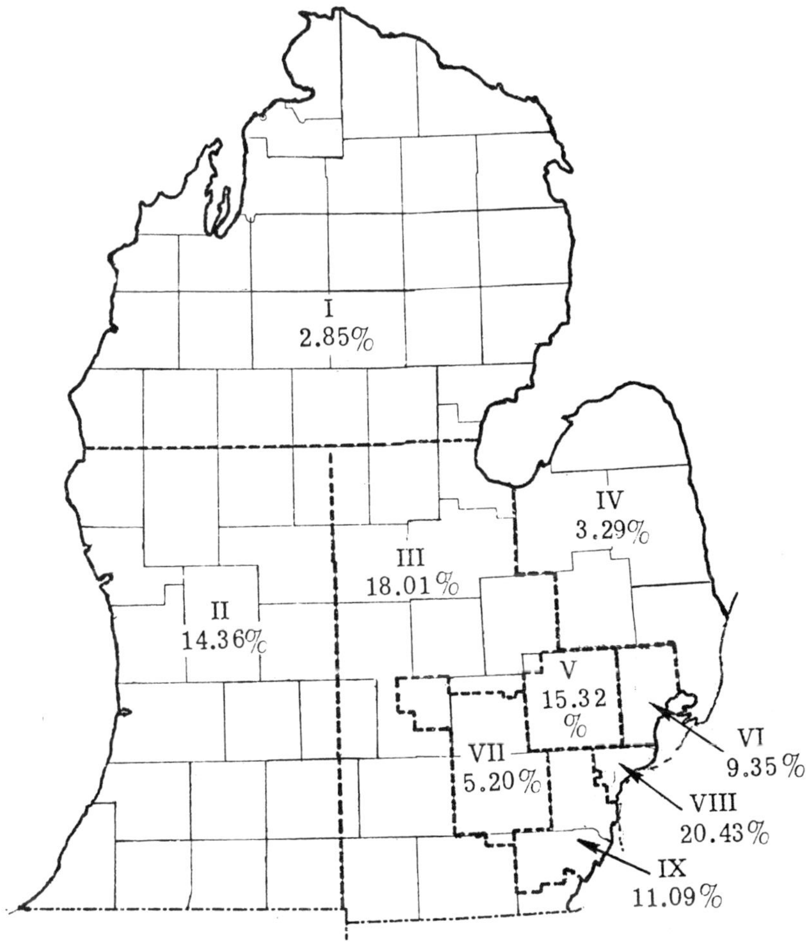

Figure 7. Load areas and present power consumption for the Detroit Edison and Consumers Power load areas.

percentages remain constant through the year 2000, the power which the completed center must supply to each area is shown in Table VII. Although the Michigan study team examined the emerging UHV transmission systems, they also concluded that these systems would not match the center growth as well as more conventional 765-kV lines. The team noted that a 1,100-kV line, with its surge impedance loading of 4,889 MWe, would only be partially loaded during early stages of the center

Table VII. Electrical Power Required in the Year 2000
by the Load Areas in Figure 7

Area Number	Load (MWe)
I	642
II	3,234
III	4,056
IV	743
V	3,459
VI	2,117
VII	1,171
VIII	4,612
IX	2,495

growth, yet at least two such lines are required to supply a given load area to insure system reliability in the event of line outages. The study also concluded that an underloaded transmission line usually requires shunt compensation to offset line charging and thus limit the steady-state line voltages to acceptable levels. In view of these considerations and the fact that the major load areas (V, VI, VII and IX) are within 100 to 125 miles of the study site, 765-kV lines appear to be more suitable for power transmission in this case. From Figure 7 it can be seen that compensated 765-kV lines supplying the load areas III through IX may be loaded to approximately twice the surge impedance loading, or 4,400 MWe per line. Because of the distances involved for load areas I and II, uncompensated lines supplying these areas cannot be loaded much above 3,000 MWe per line. Using these vlaues for line loading, the study concluded that a minimum of six 765-kV lines are required just to handle the 23,000 MWe of generating capacity. In designing an actual transmission network, twice as many 765-kV lines would be incorporated to allow several entry points into high-power density areas and improve system reliability.

The alternate Lake Michigan site chosen for consideration by the study team has the disadvantage of being separated from the major load areas by 169 to 200 miles. Since load areas V through IX have a total projected demand of 13,854 MWe by the year 2000, a minimum of five 765-kV lines would be required to satisfy these load areas alone, ignoring additional lines required for reliability. Since this site will also require three lines to service load areas I, II and III, it was concluded that the total line lengths would be substantially greater than for the first site. Cost estimates for installing 765-kV lines, including right-of-way costs, were approximately $600,000 per mile. Such additional line costs for this site would place it at an economic disadvantage unless other factors outweigh these costs.

In summary, the Michigan energy center study team suggested that a transmission network consisting of 765-kV lines was more compatible with energy center growth than UHV systems and that a network of approximately 12 such lines would supply the major load areas from the eastern site.

TRANSMISSION CONSIDERATIONS AT THE HANFORD NUCLEAR ENERGY CENTER

The electric generating capacity proposed for the Hanford Nuclear Energy Center (HNEC)[16] and the growth of near and distant loads postulated for study purposes for the year 1990 are given below:

Generating Capacity at HNEC	35,000 MWe
Load	
Onsite	8,500 MWe
Transmitted to western load centers	16,000 MWe
Transmitted to load points to the south	10,500 MWe
Total	35,000 MWe

A 16,000-MWe peak load component from the HNEC and part of the surrounding 19,500-MWe hydroelectric generation would be transmitted across the Cascade Mountain Range to load centers in Seattle, Portland and the Willamette Valley. This combined flow must be accommodated in the mountain pass corridors where bottlenecks are more likely to occur than the corridors carrying HNEC elsewhere. It was projected that the capacity in the east-west corridors must be increased about fourfold over existing capacity to accommodate the anticipated increase in power flow. This new capacity must be built in corridors radiating from Hanford to the major tie points in the present system. A major objective in the Hanford study in expanding the capacity of the system was to confine new transmission facilities to existing corridors. Although existing lines and equipment are currently of 500 kV, studies have indicated that the necessary capacity can be provided in the existing corridors with the higher-capacity overhead AC lines of up to 1,100 kV. Additional multiple circuits may be needed in certain corridors to obtain the necessary degree of system reliability, it was reported that space is available for them. Several combinations of multiple lines of various capacities were examined and it was determined that the needed expansion could be compatible with the existing transmission system in the Pacific Northwest up to 1990-1995.

It was noted in the Hanford study that at present, HVDC is economically competitive with AC transmission only for distances much greater than 200 miles. For this reason it was not deemed suitable for the east-west transmission, but might be competitive with AC for long transmission lines to California, should that be desired. It was pointed out that underground cable systems might be available for short runs to avoid congestion, improve reliability or penetrate urban areas. Also, by 1990 development of cryogenic and other advanced types of transmission might have progressed to the point where their technical and economic feasibility would suggest usage at Hanford in later years.

A preliminary estimate was made of the added transmission costs associated with the 16,000-MWe component of HNEC generation in 1990, to be sent to western load centers, relative to those that would be incurred if equal generating capacity were distributed in twin-unit stations in the general vicinity of the Puget Sound, Portland, and Willamette Valley load centers. The major assumptions made were[17]:

For the HNEC Case
- Power would be transmitted an average of 196 miles over 1,100-kV lines.
- An allowance for the cost of lower-voltage lines removed to make room for new lines is included.
- An allowance for the cost of intermediate substations required because of replacement of lower-voltage lines is included.
- Losses are included in the cost estimate for transmission.
- Transmission costs for the hydro component of power will neither increase nor decrease because of the transition to higher capacity.
- Normal peak loading on the lines will average 60% of nominal capacity.

For the Case of Distributed Generating Plants
- It was assumed that dual 500-kV single circuits from each twin-unit station would be 50 miles in length.
- Additional right-of-way for all lines would have to be acquired.
- Normal peak loading on the lines would average 70% of nominal capacity.

The cost penalty in the estimates for generation at HNEC was from $30 to $40 million/yr, or from 0.3 to 0.4 mil/kW-hr of energy transmitted. There would be an additional cost penalty against HNEC for power sent to interregional points to the south. These estimates were given in 1973 costs and require revision to today's terms.

The Hanford study noted that the occurrence of an earthquake, tornado or other storm could cause simultaneous failure of some east-west transmission lines and would have serious consequences such as shedding, for all but the most essential loads at the western load centers for a substantive period of time. The risk to service interruptions at western load centers caused by multiple transmission-line failures would be no greater than it is today, but was estimated to be greater than if dispersed siting were used west of the Cascade Range. The incidence and severity of earthquakes and tornados in this area is very low and it was observed that no storm has ever caused transmission-line outages in more than one corridor at a time.

SUMMARY OF TRANSMISSION CONSIDERATIONS

Most investigators have concluded that the cost of power transmission for energy centers would be greater than for dispersed sites; these cost penalties range from 0.2 to 0.5 mil/kW-hr. Generally there was optimism that these transmission cost penalties could be more than recovered by other savings accrued from the energy center concept. In addition, for longer-length transmission lines it may be possible to reduce these penalties by utilizing HVDC transmission.

System reliability, the property of providing continuous power, has been considered in some detail in most studies. The common opinion is that system reliability might be jeopardized by the energy center concept, but that by proper design and the use of redundant transmission lines, system reliability of energy centers could be made equal to the reliability of dispersed siting. The main concern was for so-called "common-mode" failure, occurrences such as explosions, strikes, military actions and natural disasters which can down some or all transmission lines at the same time. For this reason all energy centers are designed so that the generating units are arranged in several groups and the transmission lines disperse rapidly on leaving the vicinity of the center.

System stability will probably be more difficult to maintain at an energy center because of the longer transmission distances. Most studies concluded that stability could be maintained as well as with dispersed sites but that faster, more sophisticated switching capability may be required. In addition, HVDC capabilities might be necessary when large amounts of power need to be shifted rapidly.

Energy centers, then, have special transmission problems compared to distributed power generation, but none so serious as to argue against the concept.

V

INDUSTRIAL COLLOCATION

The most important points in the rationale behind the energy center concept are:

1. to provide a substantial proportion of new power-generating facilities to meet the future demand of a region,
2. to attempt to minimize social and environmental impacts through concentration of effects,
3. to make the most efficient use of fuel possible by providing process steam to industry and by integrating energy requirements of the center's city, and
4. to utilize as much as possible the waste heat inherently resulting from electric-power generation.

The last two goals may be achieved by constructing a variety of industrial users of electricity, heat, steam and by-products very near the power-generation facilities: industrial collocation. The objectives of collocating associated industry at an energy center and to assure the availability of various forms of energy for the collocated industries; to increase the efficiency and economy of energy utilization; to reduce the environmental impact of dissipation of heat or disposal of waste products; and to achieve increased economies derived from cooperation among industries and large-scale operations. Several other terms define energy-producing systems that employ secondary or multiple uses of energy (electricity, heat, steam, waste products). These include the "nuplex," coalplex, energy-agro-industrial complex, and modular-integrated utility systems.

An industrial plant which generates process steam could increase its steam output, generate electrical power and use or sell the power to an electric utility. Conversely, an electric utility could sell part of the steam from a generating plant as process steam to a neighboring industrial plant.[15] Such arrangements have distinct economic advantages, not only because one large boiler is cheaper than two small ones, but also because of a thermodynamic increase in the utilization of the heat content of the fuel. Such cogeneration or dual-purpose generation is not a new concept nor is it a technique restricted to energy centers. However, it does take on a new importance when combined with a complex of the size contemplated here because of the increased potential for optimally matching generation and usage. This potential rises from the greater number of alternative uses available at a large-scale operation.

At the beginnings of this century most industrial plants generated their own electricity and steam. At the present time, industry still generates about 15% of its electric requirements and uses 17% of the nation's total primary fuel for producing steam. There are in the U.S. then, two large steam systems operating relatively independently and largely for different purposes. Industry generates steam principally for manufacturing and heating; only about 30% of this steam generates any electric power before going to its final use. The utilities generate steam almost solely to feed turbine-generators producing electricity, and only a very small percentage of the steam is sold to industry.[16] In addition, there are great numbers of commercial and residential buildings requiring space heating which have their own fuel supply and utilization which could conceivably use steam. More will be said about this use in a following chapter.

Since there must have been "good reasons" (most probably economic) for the popularity of these cogeneration or dual-purpose steam plants with both utilities and industry at the turn of the century, what caused the definite reverse trend? According to Dow Chemical Company,[16] a number of factors contributed to the split in utility production over an extended period. Technical progress encouraged the introduction of larger generating units with all the associated economies-of-scale, especially as the fuel costs were increasing at the time of the two World Wars. In the early years, many urban utilities supplied steam for district heating (some still do, for example, the Toledo Edison Company). The condensate was not ordinarily recovered because it would have been too expensive to pump back to the utility. However, the more efficient higher-pressure boilers required a much higher-purity water which then was considered too costly to throw away.

At the same time, industrial managers were discovering adverse pressures on the desirability of generating power for their own operations. Some of these factors were[16]:

1. Corporate income tax rates encouraged the substitution of expenses for capital investments.
2. A difference in risk necessitated a higher return on an industrial investment than that required by utilities.

3. Decreasing electric rates from central power stations encouraged substituting power for steam in some situations.
4. Changes in the labor situation made it difficult to find and keep people able to handle the problems of coal-fired boilers.
5. Few industries were big enough to make effective utilization of the larger more efficient generating units.
6. Maintaining continuous and adequate electric power is very different from normal manufacturing processes and therefore many industrial managers preferred to leave power generation to the professionals.
7. The rate structure adopted by many utility companies tended to discourage industrial generation by providing large-use discounts.
8. Most utilities would insist on maintaining direct control over any generation facility for reasons of system stability. Such control might not match the objectives of the industry.

Even though cogeneration may be more economical for some industries, many of the above reasons will still discourage a widespread return to electric generation in industry. On the other hand, many of these same reasons argue for the extraction of steam from a central power station for industrial use, at the same time making use of the increased usable heat extracted from a given amount of fuel.

It is apparent that neither collocation nor cogeneration are exclusive attributes of an energy center. They can be advantageously used at dispersed sites also. It is at an energy center, however, where the maximum economical benefits from these approaches may be obtained. The large size of the center would allow economies-of-scale for steam generation as well as electric power generation while the greater number of different industries nearby would allow greater flexibility in design and the possibility of a more uniform demand of steam and electricity. Careful planning is required to integrate the industrial usage of steam and electricity into a utility system to avoid operating problems and to maximize the benefits of efficient and low-cost generation sources. The principal requirement is the predictability of the industrial use to allow rational planning and operation of the utility system.

INDUSTRIAL ENERGY SUPPLY CONSIDERATIONS

Once the industrial plant energy requirements have been defined, the task of designing and developing an appropriate energy supply system begins. An energy supply system which can satisfy the varied demands of an industrial process is critical to successful plant operations and the safety of the plant equipment and personnel. The energy supply system must provide the average power needs and normal fluctuations and, in some instances, must respond to unusual or transient needs as well.

The common arrangement for supplying industrial energy, both steam and electric, is for the company to purchase all electric power and generate all steam needs. This is normally referred to as the "base case" and is used for comparison with cogeneration and collocation. A schematic is shown in Figure 8.

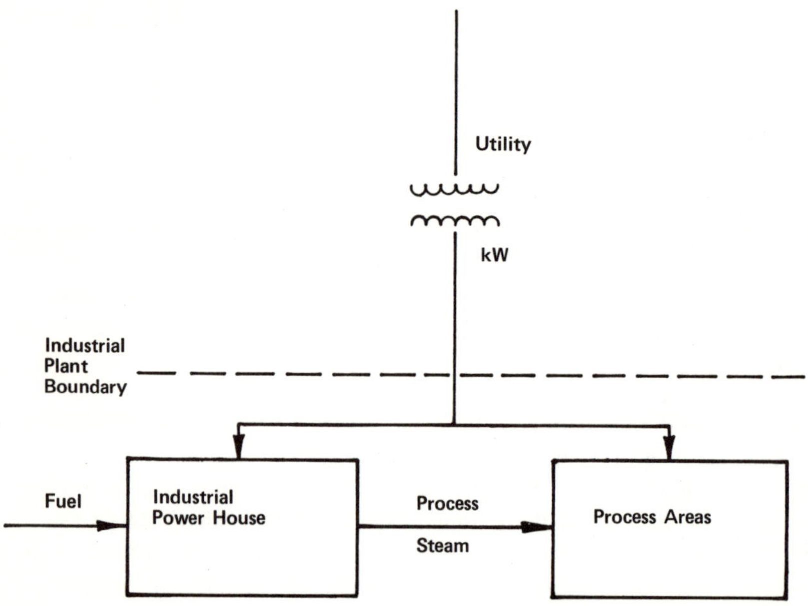

Figure 8. Base case: industrial plant without power generation. (Note: Industrial Power House supplies all process steam; Utility supplies all electric power.)

By way of comparison, industrial power generation (IPG), common in the past, is shown schematically in Figure 9. By-product electric power is generated at favorable heat rates, thus providing some industrial users with a profitable investment. These are users who need large amounts of low-pressure low-temperature steam but who have high-pressure high-temperature capabilities. The third arrangement, shown in Figure 10, would be to have a utility company devoted to the production of both steam and electric power for a group of industrials, grouped closely together, usually termed collocation. In this arrangement, there would be several expected advantages. First, the industrial plant would be relieved of the burden of generating its own process steam. Second, there would be economies realized for the industry if the system were properly designed. And, also, the efficiency of energy utilization might exceed that expected if energy supply systems were developed at each industrial plant. Here we have an

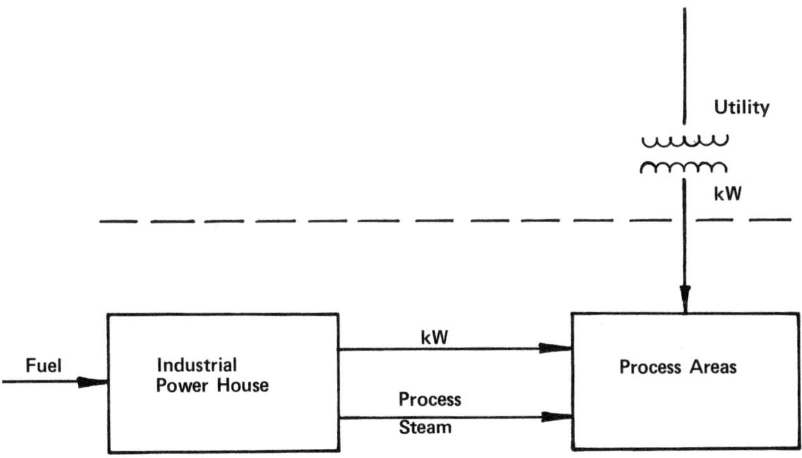

Figure 9. Industrial plant with power generation. (Note: The industry supplies all the process steam and some of the electric power.)

attractive situation where the utility can increase its yield on investment at the same time the industry realizes lower energy costs. One has the classic example of the whole achieving more than the sum of the parts, assuming, of course, that the total system is designed to operate in a near-optimum mode.

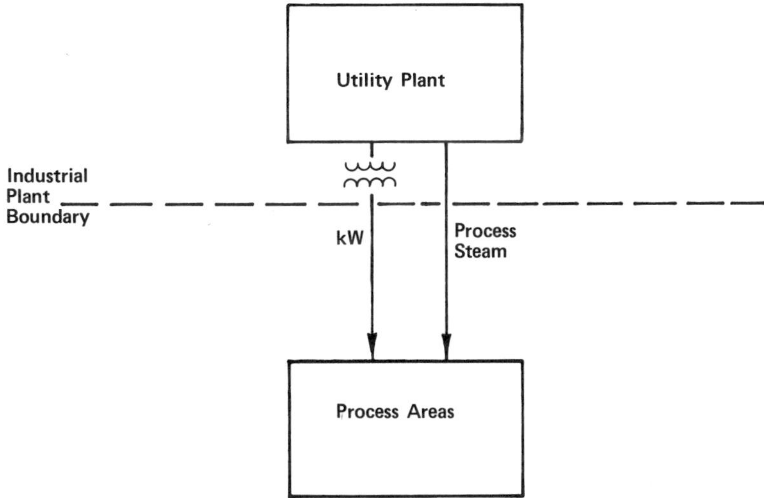

Figure 10. Collocated industrial plant. (Note: Utility supplies all process steam and electric power.)

Collocation, dual-generation and energy centers are at once separate concepts and at the same time bound closely together. Cogeneration, the production of electric power and steam by the same system, does not necessarily imply a large-scale energy center. It can be a profitable venture even for relatively small power plants. The dual-purpose central power station does imply collocation since, in distinction to electric power, steam cannot be economically transported more than a few miles. Niether collocation or cogeneration implies the need for a large-scale concentration of power generation and usage. It is, however, in such an application that the benefits of both can reach their greatest potential.

ECONOMIC ADVANTAGES OF DUAL-PURPOSE CENTRAL POWER STATIONS

Since economic advantage can be achieve by cogeneration independent of an energy center, it is possible to estimate the gains by considering a single large generating unit. The economic advantages of the energy center concept can then be estimated as a separate item.

"Dual-purpose" means a central power station, of the size and type typically constructed by utilities, which generates electric power and also furnishes a significant amount of steam to one or more customers. A number of examples of cogeneration exist in this country, the oldest being Lousiana Station No. 1 of the Gulf States Utilities Company near North Baton Rouge. The original plant was completed in 1930 and has been expanded several times; it now generates 5 million pounds per hour of steam and 240 MWe of electricity. The process steam services both the Exxon Corporation and the Ethyl Corporation. The Public Service Electric and Gas Company of New Jersey also supplies steam from their Linden station to an Exxon refinery. Between one and two million pounds per hour of steam at 150 psi are furnished from extraction turbines and sent over a mile. Consumers Power Company is constructing their first dual-purpose nuclear facility at Midland, Michigan, to furnish process steam to the Dow Chemical Company, as well as power to both Dow and the Consumers Power system. This plant is designed to supply over 4 million pounds per hour of process steam and generate 1,300 MWe of electric power. The process steam actually comes from one-half of the plant, which is also generating 500 MWe. There has been a slowdown in the construction schedule as a result of difficulties, some related to the controversy over sharing cost overruns. A brief review of the difficulties at the Midland dual-purpose plant will be presented in a later chapter.

The basic advantage of the dual-purpose central power station is illustrated in a simplified manner in Figure 11.[15] In the "power-only" case, energy must be added in the boiler to the feedwater in amounts sufficient to bring it up to point A. However, the condensing turbines which drive the generators can only utilize the amount of energy between points A and C to make electricity, and the large amounts of energy from point C back down to the feedwater level must be rejected to the environment. In the "power and steam" case, the energy from A to B is used to make electricity; the energy from D down to the feedwater level is rejected to the environment. In essence, then, the advantage lies in the fact that the industry can use much of the heat which the utility would otherwise throw away.[16] Note, however, that this is not a scheme for getting something for nothing. Any steam extracted for process steam is no longer available to generate electricity, as noted in the figure.

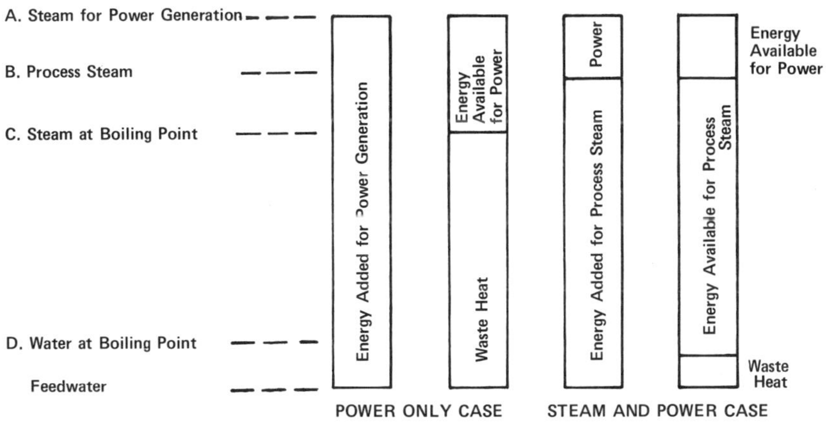

Figure 11. Comparison of useful and waste energy for alternate processes.

A better comparison of an all-electric or power-only station to dual-purpose station is obtained by working through a simple numerical example. To do this we must first introduce some elementary principles of thermodynamics. In the power-only case, the heat value of the fuel is converted into work capability in the form of electric power and is limited by the second law of thermodynamics:

$$\eta = \frac{T_s - T_r}{T_s}$$

where T_s is the absolute temperature of the input steam to the turbine, and T_s is the temperature of the heat sink, usually the atmosphere. This law limits the efficiency with which heat can be converted into work. In chapter III, mention was made of a standard fossil-fueled generating system of 885 MWe with a turbine inlet temperature of 1,000°F. If this turbine has an exit temperature of 100°F, then the maximum theoretical efficiency is (1460-560)/1,460 or about 61%. The temperatures have been converted to absolute as required by the expression. This maximum possible is never obtained because of pumping requirements, frictional losses and heat losses. In fact, it turns out that nearly two-thirds the heat content of the fuel is rejected to the environment in a power-only station.

Many industrial plants have, in addition to a power load, requirements for heat to be used for drying, cooking, heating products or heating buildings. If the heat being rejected to the environment by the power plant can be used in the industrial processes, then nearly all the heat content of the fuel can be put to use. Unfortunately, in modern power plants with condensing turbines this heat, while plentiful, is of such low temperature (85-100°F) as to be useless. If a portion of the steam is extracted at some point between the stages of the turbine it will have the temperature and pressure required for industrial processes. Although this extracted steam no longer is available to work in the turbine, and the power output of the turbine is less, more of the fuel's heat content is useful and less heat needs to be rejected to the environment. This case is best illustrated by the numerical example that follows.[17] Suppose an industry requires 5,000 kW of power and 45,000,000 Btu/hr of heating load. Consider two systems for providing these requirements. One system consists of two boilers—one for generating just process steam, the other for power only; the second system provides both power and steam by use of an automatic extraction turbine. The first system is shown in Figure 12 and consists of two boilers, a process heater, a turbine, condenser and the necessary condensate and feedwater pumps. The enthalpies at various points in the flow diagram of the first plant are:

$$h_1 = 1,363 \text{ Btu/lb} \qquad h_4 = 936$$
$$h_2 = 1,363 \qquad\qquad h_5 = 83$$
$$h_3 = 208$$

The heat liberated by one lb of steam passing through the process heater is

$$Q = 1,363 - 208 = 1,155 \text{ Btu/lb}$$

Hence, the rate of flow required is

$$\frac{45,000,000}{1155} = 39,000 \text{ lb/hr}$$

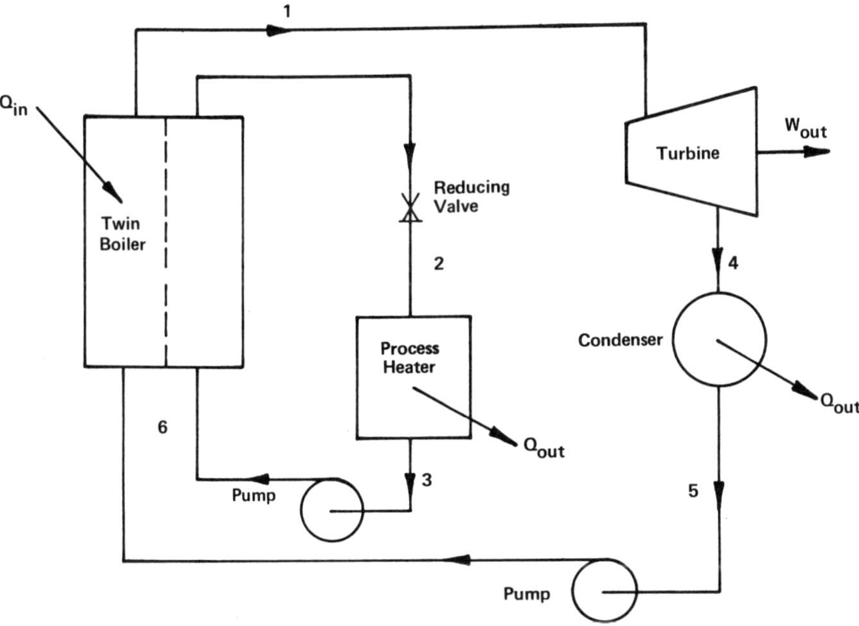

Figure 12. Flow diagram of power and heating plant. (From *Basic Thermodynamics* by Charles Leonard Brown. Copyright 1951. Used with permission of McGraw-Hill Book Company.)

The work done by a pound of steam in the turbine is

$$W = 1,363 - 936 = 427 \text{ Btu/lb}$$

The steam rate is

$$R = 3,413/427 = 7.98 \text{ lb/kWh}$$

Therefore, the mass rate of flow through the turbine to produce 5,000 kW is

$$M = 5,000 \times 7.98 = 39,900 \text{ lb/hr}$$

Since the heating and power flows are separate, the total rate of steam flow is

$$M = 39,000 + 39,900 = 78,900 \text{ lb/hr}$$

The heat rejected by the condenser to the environment per unit mass of steam is

$$Q = 936 - 83 = 853 \text{ Btu/lb}$$

and the total heat rejected is

$$Q = 853 \times 39{,}900 = 34{,}100{,}000 \text{ Btu/hr}$$

Next an energy balance can be applied to the boiler to determine the heat supplied.

$$Q = 78{,}900 \times 1{,}363 - 39{,}000 \times 208 - 39{,}900 \times 83 = 96{,}100{,}000 \text{ Btu/hr}$$

It should be noted that the two boilers diagrammed in Figure 12 have been positioned side-by-side to suggest that the steam rate could have been supplied by a single larger boiler. This would be the preferred approach since the larger boiler would be cheaper and have an overall efficiency greater than two separate ones. Nevertheless, this is not the only gain achieved by a dual-purpose central station.

The flow diagram of the second plant, using automatic extraction, is shown in Figure 13, and the associated enthalpies are:

$$\begin{aligned}
h_1 &= 1{,}363 & h_4 &= 936 \\
h_2 &= 1{,}109 & h_5 &= 83 \\
h_3 &= 208 &&
\end{aligned}$$

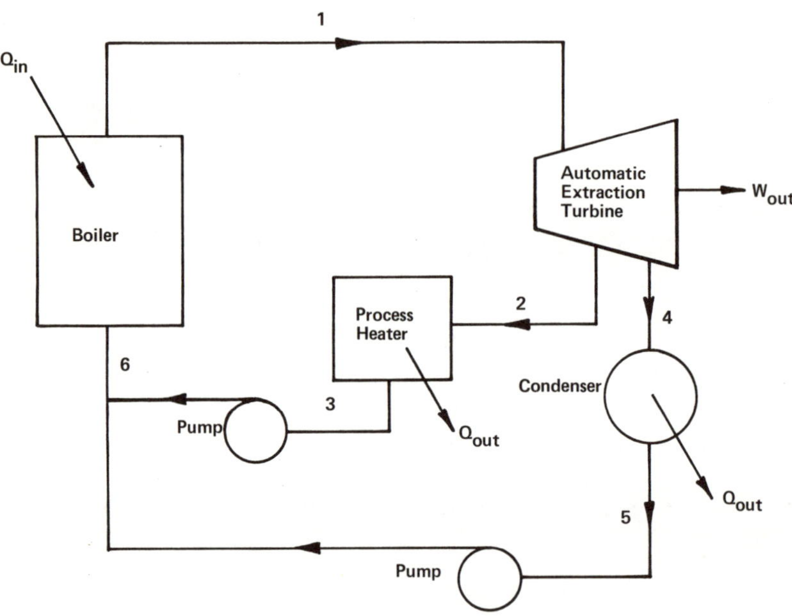

Figure 13. Flow diagram of a power and heating plant where the heating system is provided by an extraction turbine.(From *Basic Thermodynamics* by Charles Leonard Brown. Copyright 1951. Used with permission of McGraw-Hill Book Company.)

The heat given off by steam condensing in the heater is
$$Q = 1,109 - 208 = 901 \text{ Btu/lb}$$

and the rate of flow required is

$$M = \frac{45,000,000}{901} = 49,900 \text{ lb/hr}$$

However, this steam expands partially in the turbine doing work of

$$W = 1,363 - 1,109 = 254 \text{ Btu/lb}$$

Consequently, the power produced by the steam bled from the turbine is

$$\frac{49,900 \times 254}{3,413} = 4,460 \text{ kW}$$

An additional 5,000 - 4,460 = 540 kW must be generated. This is done by admitting more steam to the turbine and expanding it completely to the condenser. The steam rate for this operation is the same computed for the first example; 7.98 lb/kWh. Therefore, the additional steam required is

$$540 \times 7.98 = 5,300 \text{ lb/hr}$$

The total steam flow is

$$M = 49,900 + 5,300 = 55,200 \text{ lb/hr}$$

Since now only 5,300 lb/hr passes into the condenser, the heat rejected there is

$$Q = 5,300(936 - 83) = 4,520,000 \text{ Btu/hr}$$

The heat input to the boiler is

$$Q = 55,200 \times 1,363 - 49,900 \times 208 - 5,300 \times 83$$
$$= 64,400,000 \text{ Btu/hr}$$

In comparing these two examples, one sees that when process steam is extracted, the boiler needs to supply less steam (here 33% less) and there is a very great reduction in waste heat rejected to the environment. This saving is not obtained without problems. First, it is not economical to transport steam more than a few miles because of the pressure. Second, there is the problem of what to do with the condensate. It may be costly to return it to the boiler but it is also expensive to throw away the the high purity water demanded by modern high-pressure high-temperature boilers. In the above examples, pumping and feeding makeup costs have been neglected and these would reduce the gains achievable.

For a practical dual-purpose generating plant at an energy center, the exact savings in fuel and capital costs would depend on many factors related to the requirements for process steam, engineering design of the overall system and whether the condensate should be returned. It is probably desirable to make use of existing standard units for steam and electricity generation rather than develop new special purpose systems for specific requirements. Since only certain steam pressures and temperatures would be economically extracted from standard units, it will be necessary to very carefully select those industries which can utilize the steam available at the usual extraction points.

One economic study[13] was performed on the collocation of industries using process steam at an energy center. The conclusions reached were that for most industrial users of process steam, it may be more economical for the industry to produce its own steam than to purchase steam from a central power plant. However, the conclusions reached on this issue would depend on the conditions at a particular installation with respect to magnitude of individual process steam requirements, distance between power plant and industry, use of by-product fuels, and provisions for space capacity. Another consideration would be the attitude and policies of the management of the industrial concern. The study further states that some industries will find it advantageous to purchase rather than produce all power and steam requirements. This decision would minimize the requirements for technical management, operating experience and capital investment in an area not of primary concern to management. For the utilities, such an arrangement would represent an additional source of revenue in a business area of primary interest.

Since the operating life of power plants is long compared to that of many industrial plants (30 years versus 10 or 12), long contract lives might cause concern to some potential energy center participants. This problem could be minimized at an energy center, however, because of the greater number of possible users, both old and new. Also, the power plant would have some flexibility in allocating steam for process use or power generation.

OVERVIEW OF ASSOCIATED INDUSTRY

The main advantage of collocation and dual-purpose central stations has been shown to be that industries can make good use of much of the heat utilities would otherwise throw away. In addition, the associated industries could achieve a variety of objectives, including the assured availability of various forms of energy for collocated industries, increased efficiency and economy of energy utilization, reduced environmental impact from heat dissipation or waste products disposal and increased economies derived from a balanced cooperation among industries. However, these objectives can only be met by careful planning and the optimization of steam, power and central power station use. This means identifying the appropriate industries that could be located at an energy center and adjusting outputs and inputs of the various elements. Most studies of energy centers have identified candidate industries including:

1. Coal gasification and liquefaction facilities that would produce low-Btu gas or synthetic fuel oil needed by center industries to replace increasingly scarce fuels. The coal-conversion facilities would convert the coal, which may contain large amounts of sulfur, to clean-burning fuels.

2. Industries that engage in the economical production of specific chemical products used by other industries at the center such as hydrogen, oxygen or chlorine.

3. Industries that use large amounts of process steam. These industries would take advantage of the increased fuel efficiency inherent in the dual-generation of electricity and process steam as described in the previous section. In 1968, energy used for process steam (exclusive of power generation) constituted 52% of total industrial use of energy.[16]

4. Industries that require large amounts of natural gas. The special clean-burning characteristics of gas are required for certain industrial processes such as glass manufacture and some metal processing. Low-Btu gas, available from the collocated coal gasification plants, would be in great demand at an energy center because it is adaptable and nearly pollution-free.

5. Agricultural and aquacultural enterprises capable of using the large amounts of low-quality waste heat that would otherwise be rejected to the environment. The heat used in this manner is a beneficial by-product of the electric generation process and would reduce the need for some costly cooling towers or cooling ponds. Specific applications of this use of waste heat will be described at the end of this chapter.

6. Industries that are devoted to the production or use of mineral resources located within economic shipping distance of the center. These potentially appropriate industries would naturally be determined by the location of the center. In Michigan, for example, the large reserves of iron ore and limestone would suggest steel and cement mills. In the Wasatch Front area of Utah, the availability of Utah crude oil and Great Salt Lake brine suggests the manufacture of chlorine and chlorinated hydrocarbons.

7. Industries that make use of natural resources such as forests and timber which may be readily available at some center sites. Such industries would include pulp and paper manufacture, wood furniture and lumber products.

8. Industries that would provide materials for the center's construction-such as cement or steel reinforcing bar. Reinforcing bar would be one of the products supplied by a mini-steel mill which requires electric power.

Possible Collocated Industrial Configurations

Each of the possible energy center site locations, whether Utah, Oklahoma, Michigan or elsewhere, has its own characteristics which dictate what industries could be economically collocated there. These characteristics include the availability of water, mineral resources, natural resources, transportation facilities and markets. Industrial configurations selected for consideration at some energy center sites are described in the following sections.

Michigan Industrial Configuration

A Michigan energy center is a logical location for coal gasification facilities. The low-Btu gas (up to 300 Btu/ft^3) can be used for many of the collocated industrial processes. It should be noted that although low-Btu coal gasification processes are less complex and more efficient than the processes for producing high-Btu gas, the low-Btu gas has such a low energy density that it is economically prohibitive to transport over large distances. Thus, low-Btu gasification should take place near the point of end use; the energy center concept meets this requirement. Low-Btu gas would be directly useful in many industrial processes which now depend on oil or natural gas as primary energy sources; it would be a natural complement to electric energy and process steam for these industries.

An attractive method of utilizing the coal gasification process is to use the gas in a combined-cycle system with both gas and steam turbines producing electric power. The necessary technology is under active development and testing at this time. Since the gas is to be used to provide a variable power-plant load, the gasification process must be near a large-scale gas storage facility. This option seems particularly attractive for Michigan as many underground storage sites might be available. Michigan's gas storage facilities are already the largest in the U.S. with 781 billion ft^3, 13% of the national storage capacity.[18]

A pulp and paper industry would probably be attracted to a Michigan energy center site as a result of the assured availability of electric power and low-cost process steam and by the availability of good transportation for the nearby raw materials. Certain chemicals used in paper manufacture, such as chlorine and sulfur, could be available from other Michigan energy center activities.

Location of a petroleum refinery at a Michigan energy center would offer many favorable advantages including the availability of fuel, water transportation and distribution facilities, labor supply and, of course, a suitable site. Petrochemicals from the refinery would also be available as feedstock for other center industries.

The Michigan energy center study team expected chemical industries to be an important component of their concept. Without precisely defining a specific group of chemical products, the Michigan team outlined the major characteristics of such a chemical complex. The chemical complex in Michigan could be based on a combination of raw materials abundant in the Great Lakes area. Depending on the availability of raw materials, markets for the resulting products and the economics of producing them at the Michigan site, these processes might use coal, petroleum, timber, salt, limestone and metal ores within economical transportation distance to the site. Other raw materials would be derived from industries at the center. These would include hydrocarbons and aromatics from coal liquefaction and sulfur from the now-required desulfurization of coal.

As an alternative to the construction of a full-scale steel mill using iron ore as its raw material, the Michigan study suggested the construction of a mini-mill with an investment cost only 25% that of a large plant. Although products from such a mini-mill would vary with location and market, they usually would include concrete reinforcement bars which could be used right at the energy center during later phases of its construction. Transportation costs for raw materials and finished goods are major expenses for the steel industry. Location of mini-mills near the customer gives the mill a significant competitive advantage over more distant producers. In addition, Michigan has a scrap surplus, a sizable fraction of which could be consumed in the construction of the center itself. Many of the mini-mills are designed to use electric-arc furnaces; a fundamental requirement for successful operation is a guaranteed supply of electricity at an attractive cost. The mini-mill seems to fit very well into the Michigan energy center concept.

The Wasatch Front Industrial Configuration

The Dow Chemical Company, in a 1973 study,[5] devised an industrial complex suited to the Wasatch Front Area near Salt Lake City, Utah. From a long list of potential end products, Dow selected an appropriate mix by informally applying considerations such as profitability, demand, availability of raw materials, distribution costs, available technology and compatibility. The availability of brine, Utah crude and coal were the determining raw materials. Out of some 34 products, 10 survived the winnowing process: propylene, caustic soda, magnesium, low-density polyethylene, high-density polyethylene, propylene oxide, vinyl chloride, perchloroethylene, carbon tetrachloride and methyl chloroform. The relationships among the raw materials and the intermediate and final products is shown in Figure 14. The utilities required for this industrial complex are approximately 310 MWe of electric power, 2,100,000 pounds of steam per hour, 6,000,000 gpd of desalted water, and 20,000,000 gpd of brackish water. Approximately 5,000 acres of land and nearly $500,000,000 of investment capital would be required. The complex is estimated to generate $275,000,000 annually and to provide employment for more than 1,000 people.

In this initial study by Dow, it was concluded that such an industrial chemical complex could be constructed near Salt Lake City which would utilize Utah raw materials, provide additional employment, furnish an industrial base for further development and yield adequate profits.

Earlier in this chapter, the fuel savings and increased efficiency of dual-purpose generation was discussed. However, the selling price of process steam was not computed because of the variety of ways that the benefits of cogeneration can be divided between the power company and the consumer. It is instructive to examine the procedure that Dow Chemical Company used to determine the price for steam. They assumed an equal sharing of the benefit of a dual-purpose plant between the utility and customer. The steam load of 2,100,000 lb/hr would reduce power production by 100 MWe. Including the power load, the industrial complex would be utilizing essentially all the capability of one power plant (assumed to be sized at 400 MWe). Dow assumed that the complex would operate at 95% capacity and the cost of operating one power plant unit at this load factor would be $39,760,000 per year. The benefit was found by comparing this figure with the cost of purchasing 300 MWe of electricity at 12.84 mil/kWh and generating 2,100,000 pounds of steam per hour in oil-fired boilers. The cost of this power would be $32,060,000 per year and the cost of operating the boiler was estimated to be $23,000,000 per year, for a total of $55,060,000 per year. The total benefit of going to a dual-purpose plant is thus the difference of $15,300,000 per year. Dow then subtracted half of this from the cost of operating the boiler and arrived at an annual purchased steam cost of $15,350,000. For the annual steam consumption required, this results in a price for steam of $0.88 per thousand pounds.

Industrial Configuration at the
Puerto Rico Energy Center

The Puerto Rico energy center is much smaller (less than 3,000 MWe) than those proposed for other sites, but is otherwise similar, such as in the use of cogeneration, collocation and the desire to match several industrial processes together for conservation of energy and materials. It is therefore not surprising that elements for inclusion in the industrial

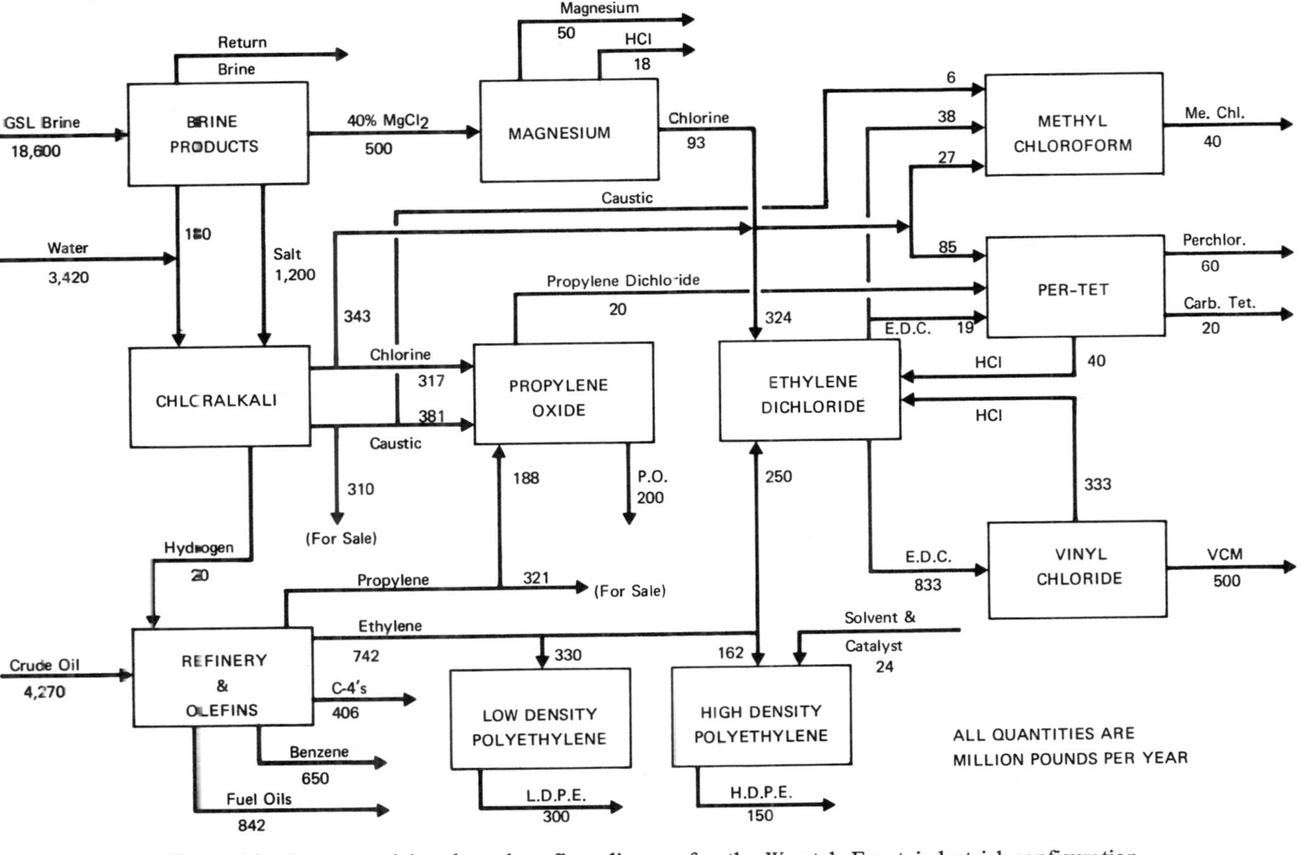

Figure 14. Raw material and product flow diagram for the Wasatch Front industrial configuration.

segment were tested against such criteria as market positions for the products, profitability, competition, proximity to raw materials and distribution costs. In addition, potential product elements were considered in view of several advantages that Puerto Rico offers including a possibly favorable oil quota position, U.S. tariff protection, a tax incentive plan and a deep harbor. The Puerto Rico study[4] also mentioned the disadvantages of higher distribution costs, lack of a large local industrial market, and the absence of local mineral deposits except for copper ore. A candidate list of plants and products used by the study included:

Chemical Refinery	Silicon	Phosphoric Acid
Vinychloride	Propylene Glycol	Copper Refining
Magnesium Hydroxide	Heavy Water	Chlorine-Caustic
Ethylene	Mini-Steel Mill	Fuel Reprocessing
Aluminum	Ethylene Dichloride	Hydrogen
Ethylbenzene	Ammonia	Hypochlorite
Magnesium	Cumene	

The economics of the potential products were screened and the surviving products were sized so that the raw material requirements balanced; the plants were compatible with projected technology and the plant capacity would make good business sense for the 1980-1990 time period. Although the economic viability of each module was examined, feasibility was ultimately judged on the industrial segment as a whole and consideration of the economic advantage of complexing, the major premise behind every energy center. After examining several cases of complexing and using the criteria, advantages and disadvantages outlined above, the configuration selected consists of a 60,000-bbl/day refinery, a chlorine-caustic plant of 1,470 ton/day, and the production of ethylene dichloride, ethylene glycol and high-density polyethylene. The estimated capital cost was $468.2 million in 1970. A schematic diagram of the proposed complex, showing the interconnected flow of products, is given in Figure Figure 15.

Industrial Facilities For Camp Gruber

At Camp Gruber, as with the other proposed energy center sites, it is the special local characteristics which determine what industrial configurations are feasible. The single most important favorable factor in this part of Oklahoma is the availability of land. The Camp Gruber Military Reservation is a former 65,000-ac U.S. Army training camp established in 1941. About half this land has been transferred to the state of Oklahoma and the rest has been declared "surplus" by the federal government. It is possible that the entire reservation could be transferred to the state.

The Camp Gruber study considered using some of this land for an energy center. Land availability was an important factor in selecting compatible industries for Camp Gruber.[2]

Large-scale conversion of coal to fuel gases and liquid fuels requires large facilities and would thus be particularly suited to siting at Camp Gruber. For the Camp Gruber study, a coal gasification facility was assumed to consist of a large (20,000 to 40,000 tons of coal per day) plant designed to provide a gas with the same heat content as natural gas. The gas product would be shipped offsite by pipeline and by-product facilities would produce sulfur and various oils.

Since the process details of the several existing coal gasification processes have similar siting requirements, the HYGAS process was used as the example design in the Camp Gruber study. In the HYGAS process, pulverized dried coal is fed into a reactor feed system where it is heated and mixed with process oil. This mixture is then fed into a reactor where the coal reacts with hydrogen and steam to form a variety of reaction products including the desired methane as well as several impurities, hydrogen and carbon monoxide. After removing most of the impurities, the gas enters a second reactor vessel where the carbon monoxide and hydrogen are converted into methane. The heat content of the resulting gas is similar to natural gas. Coal liquefaction processes were also considered for the Camp Gruber site. Liquefaction plants are similar to gasification plants but may require more land because of the additional facilities necessary for separating the liquid products and converting them into other chemicals or using them as the basis for other manufacturing processes.

Energy-intensive industries were considered for the Oklahoma site. These included aluminum reduction, magnesium reduction and uranium enrichment. An aluminum reduction plant converts purified aluminum ore into aluminum metal ingots in electrochemical cells. The process of reducing magnesium is similar and both use large quantities of cheap reliable power.

Uranium enrichment increases the U-235 content of uranium from the normal 0.7% to the higher concentration needed for the operation of nuclear power plants. It is estimated that at least one additional large enrichment plant will be needed in this country by 1984 to meet foreign and domestic demand. Uranium enrichment is practically achieved by either centrifugal force or diffusion. In a centrifuge plant, the U-238 is separated from the U-235 because of the higher molecular weight of U-238. In a diffusion plant, separation occurs because the smaller molecular size and greater mobility allows U-235 to pass through a barrier faster than U-238. In both plants, many individual stages are needed to obtain the desired concentration of U-235. The primary requirements

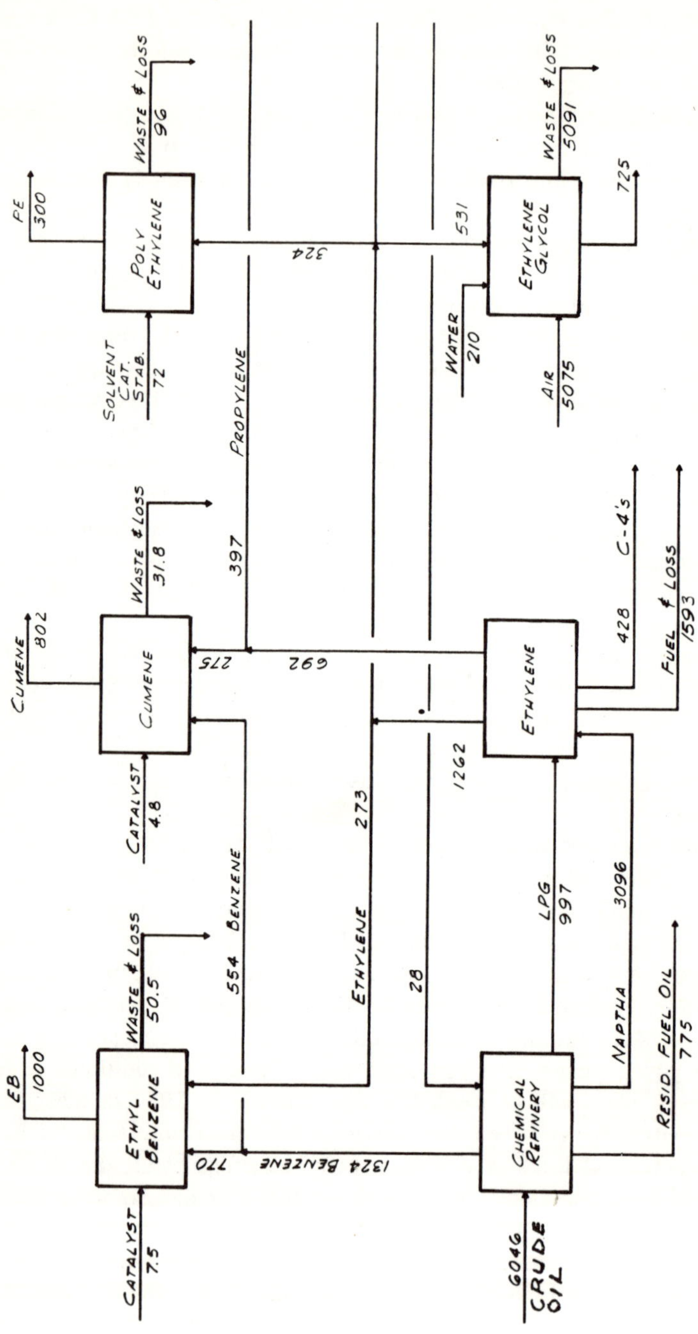
POLY ETHYLENE
PE 300
WASTE & LOSS 96
SOLVENT CAT. STAB. 72
PROPYLENE
324
CUMENE
CUMENE 802
WASTE & LOSS 31.8
397
CATALYST 4.8
275
269
692
ETHYLENE
C-4's 428
FUEL & LOSS 1593
1262
ETHYLENE 273
28
ETHYL BENZENE
EB 1000
WASTE & LOSS 50.5
BENZENE 554
CATALYST 7.5
770
1324 BENZENE
CHEMICAL REFINERY
LPG 997
NAPTHA 3096
RESID. FUEL OIL 775
CRUDE OIL 6046
ETHYLENE GLYCOL
WASTE & LOSS 5091
725
531
WATER 210
AIR 5075

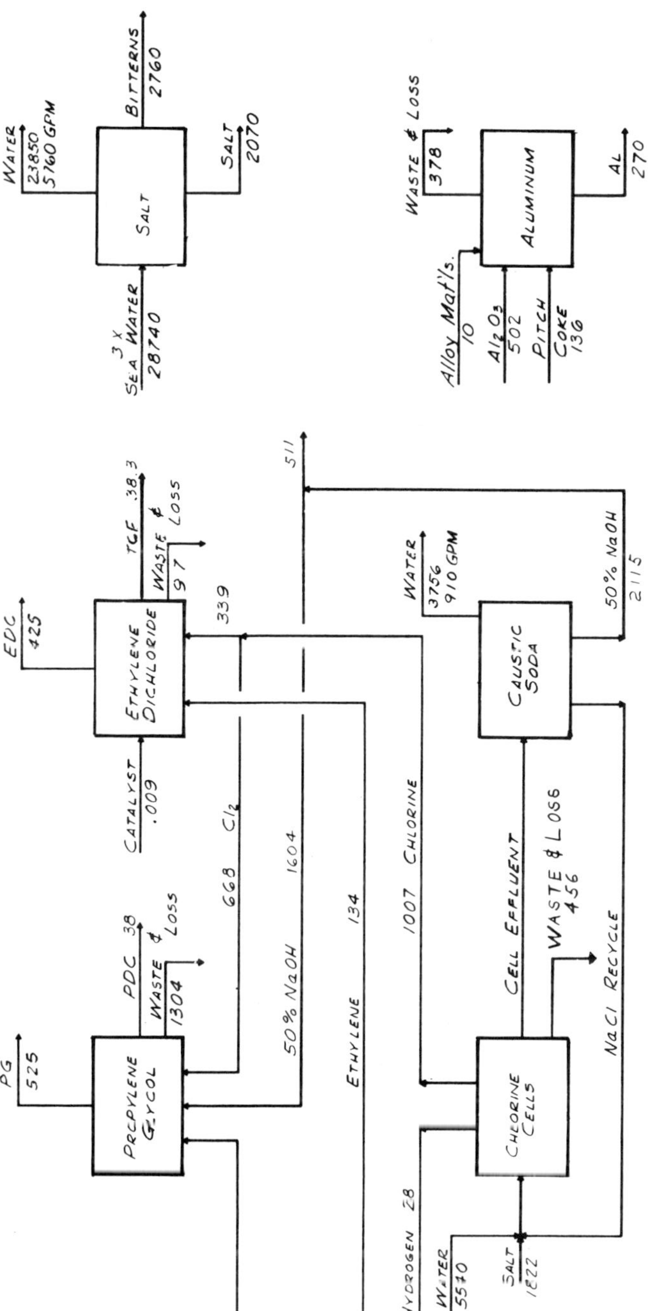

Figure 15. Schematic diagram of the proposed Puerto Rico energy center showing the interconnected flow of products.

for either type of plant are a large reliable power supply (2,000 MWe for the smallest competitive diffusion plant), and a moderate amount of land (200 to 400 ac).

The Camp Gruber study assumed a standard diffusion plant with a capacity of 8,750,000 separative work units per year and which would produce low-enrichment uranium up to 4% U-235. The feed uranium would be shipped to the plant in the form of uranium hexafluoride. The product would be shipped to fuel fabricators in the same but enriched form.

Other energy-related industries, including uranium conversion, nuclear fuel fabrication, nuclear fuel reprocessing, nuclear waste storage and petroleum refining, were also considered for inclusion at Camp Gruber. Although each industry has its own particular siting requirements, it was felt that any one of them could be located at Camp Gruber.

Summary of Industry Configurations

In the preceding sections we have seen that each projected energy site has some unique characteristic which can be exploited by the proper mix of industry types. In Michigan we have the almost unlimited supply of fresh water leading to the consideration of industries needing large quantities of water such as paper and pulp mills. In the Wasatch Front area, the ready availability of brine and petroleum led to a complex of chemical plants producing a variety of chlorinated petrochemicals. A deep water harbor in Puerto Rico suggested a manufacturing complex using readily shipped raw materials and producing products useful for the existing markets. At Camp Gruber, a large expanse of uninhabited land is a positive factor for the siting of various nuclear energy processing facilities. And, of course, at all sites the concept of a large-scale power generation capability would provide a large supply of reliable energy for those electric-power-intensive industries. In the following sections, a selection of candidate industries will be described in more detail including the input, output and siting requirements.

COAL GASIFICATION AND LIQUEFACTION

With the decreasing availability of oil and natural gas, coal will become more important over the years as a major source of energy. For many residential, industrial and transportation applications, coal cannot be used directly but must be converted to a liquid or gaseous form. We are not likely to see a return to coal-fired home stokers nor to large coal trucks rumbling through our city streets with their dusty loads. In addition, we could not tolerate the increased air pollution likely to result from the direct burning of high-sulfur coal. Since an elimination of these problems can be achieved by methods of coal gasification and liquefaction, a large research and development effort is aimed at the improvement of existing methods or development of new methods. Coal gasification and lique-faction has the added advantage of being a totally domestic industry not dependent on imported feedstocks or subject to international vagaries.

Although a large number of coal gasification and liquefaction processes might be suitable for industrial center installation, this discussion is con-fined to the general nature of processes which could be profitably located at an energy center. The advantages and requirements of siting coal processing facilities at an energy center will be described for a general class of techniques rather than for a specific process since new develop-ments in coal technology are proceeding rapidly.

Process Description

The Lurgi and Koppers-Totzek processes are the two methods of coal gasification currently available on a commercial scale. Discussions of coal gasification usually make a distinction between low-Btu and high-Btu gas. Low-Btu gas is simpler to derive from coal since high-Btu production requires an additional process. Natural gas, chiefly methane, has a heat content of 1,000 to 1,100 Btu/ft^3 while low-Btu gas, composed mainly of hydrogen and carbon monoxide, has a heat content of only about 300 Btu/ft^3. New coals under consideration as feedstocks for gasification or liquefaction have hydrogen-to-carbon ratios lower than one. This means that if the higher heat content methane is to be produced, hydrogen must be added or carbon removed, or some combination of both. The most obvious source of hydrogen is water and this is used in all processes. The reaction that one wishes to accomplish during gasification is:

$$coal + water \longrightarrow CH_4 + CO_2$$

To achieve this, crushed and dried coal is fed to a gasifier and mixed with steam and oxygen at about 400 psia and 1,300°F. The result is a low-Btu gas containing hydrogen, carbon monoxide and carbon dioxide. It also contains what at this point are considered impurities: nitrogen, sulfur compounds, unprocessed coal, ash, oils and tars. Since it is really not usable in this form, most impurities must be removed. The oil and tars are removed by a quenching process and the gas is reacted with water to increase the proportion of hydrogen by a quenching process. This pro-duces a gas with a heating value of around 300 Btu/ft^3. After shifting to a higher hydrogen content, the heat content is about the same. In this form, the gas may be very useful for certain industries which are collocated with the gas production. It is not economical to transport low-Btu gas over long distances because of the low heat content on a volume basis. It is, however, useful locally because it is unnecessary to add the further step of methanation. There are a number of schemes for changing hydrogen and carbon monoxide to methane but all introduce low-Btu gas into a reactor at 100-550°F and at a pressure up to 1,000 psia in the presence of a suitable catalyst. After reaching the temperature which initiates the reaction, the process is exothermic, that is, it gives off heat; this heat must be removed to keep the reaction within the proper operating temperatures range. Table VIII gives the approximate percentages of various constituents of the products at each stage of the process.

Table VIII. Major Constituents of Stages in the Coal Gasification Process

	Coal	Low-Btu Gas	Shift Process	High-Btu Gas
C	60%	—	—	—
H_2	40%	48	66	5
CO	—	36	3	0.2
CO_2	—	4	30	0.2
CH_4	—	1	1	92

In this table we have excluded the proportions of H_2O, N_2 and some higher hydrocarbons. Nevertheless, the change in the important components can be seen.

It was noted above that the methanation process was exothermic. It is necessary to remove this heat so that the reaction may procede at an optimum temperature and pressure. It is a highly speculative potential, but it may be possible to profitably use this heat for other processes in the energy center. For example, the heat is usually removed by heat exchangers using water as the medium of exchange. It now becomes possible to use the heated water at other points of the system where it would be needed, for example, as heated feedwater for the power unit boilers. It would be necessary to conduct a thorough economic analysis of the entire system to determine if energy savings are possible. But this potential is just one of the many advantages possible when combining a large number of industrial processes with power generating facilities. Some collocated processes give up heat, while others require heat. It is likely that optimization of the overall system would lead to several instances of give-and-take which would result in overall energy conservation. In a later chapter, when we discuss residential and transportation applications, further opportunities for the utilization and optimization of coal gasification will be discussed.

Coal Gasification at the Michigan Energy Center

Low-Btu gas would be directly useful in many industrial processes that now depend on oil or natural gas as the primary energy source such as the pulp and paper mills suggested for the Michigan center. Average energy requirements for a 2,000-ton per day pulp and paper mill are around 42 million ft^3/day of 300-Btu gas. A petroleum refinery would consume 230 million ft^3/day. Low-Btu gas would be particularly important for some industrial processes, such as steel heat treating or glass manufacture, where the clean-burning properties of gaseous fuels are important.

There are also economic incentives for using low-Btu gas for power generation. Two inherent advantages of low-Btu gas over high-Btu gas for utility use are present: the higher efficiency of converting energy in raw coal to energy in the gas (no loss in the additional methanation stage), and the lower cost per unit of energy in the gas if the gasification plant is base-loaded (possible at an energy center).

Another method of utilizing the coal gasification process is by using the gas in a combined-cycle power generator having both gas turbines and steam turbines to produce electric power. This technology is currently under active development and testing. Even with existing equipment, the low-Btu gas combined-cycle system may be at least competitive with conventional systems. With the expected improvements in gasification processes and gas turbine equipment, these systems will become even more competitive because of anticipated thermal efficiencies up to 50%. It may be possible that these new systems will require less space, have fewer cooling water requirements and offer simpler means for controlling pollution.

If the gas is used to provide a variable power plant load (that is, peak loading rather than base loading), the gasification process must either have the ability to provide gas at a variable rate or be associated with large-scale gas storage facilities. Generally speaking, gasification plants cannot be turned down more than a few percent. It may also be possible to use the gas for an industrial process load which can be varied to free-up gas during electric peak loads. An incentive for this load management function would be a reduced price on gas to the industrial user to compensate for the greater difficulties and higher costs attending the load management process. This is another instance where careful analysis of the give-and-take relationships in an energy center could show a net energy savings or economic advantage.

Long-distance transportation of low-Btu gas is not economical nor is large-scale storage. It is probably marginally economical for the large-scale storage of 300-Btu/ft^3 gas but not for gas having lower heat content. Other advantages would make it worth the extra cost for storage. For example, if low-Btu gas is used as a fuel source for electric power generation, then process reliability is very important. Because of the need for continuity of operation by utilities, the coal gasification process must show a reliability comparable to that of the generating plant, otherwise there must be a constant supply of gas from storage. This constancy of supply, available from storage, may be worth the storage cost of low-Btu gas.

There are several sites in Michigan (for example, its salt mines) where gas storage might be feasible. As stated, Michigan's gas storage facilities are already the largest in the U.S.

Whether to store the gas, whether to use gas turbines for base load or peaking or neither, whether to use gas load management for industrial users, whether to generate low- or high-Btu gas, these are all complicated and interrelated choices which affect the economics of the total system. Analysis and optimization of these choices and mixes have not yet been performed. It no doubt can be accomplished by computer simulation. But it is precisely at this point that the concept of a large-scale energy center becomes powerful. A large number of energy users and energy producers at one site can, by careful planning, be optimally matched to each other so as to achieve a net energy conservation. Although the greater number of energy producers and users certainly complicates the analysis problem, their greater number allows a flexibility that cannot be achieved any other way.

Other examples of how multiple users of energy and large-scale operations may be collocated will be described in other sections. One example may be considered, at least in a qualitative way, at this time: low-Btu gas may be uneconomical to store, high-Btu gas is more expensive to produce. It may turn out that an analysis of the optimum heat content of a gas takes on some intermediate value, say 500 Btu/ft^3. This might be provided by a large low-Btu production facility and a smaller high-Btu plant with a subsequent mixing. However, such an intermediate Btu gas might have a reduced usableness unless the user facilities were specially designed for it. Planning the entire energy center as a coordinated whole would make such choices feasible and increase the overall efficiency of fuel utilization.

Certain methods of coal gasification will result in by-products such as ammonia and sulfur, which can be used as feedstocks for many types of chemical manufacture. The assured availability of energy in various forms and chemical feedstocks for different types of industry would be an incentive to locate industry in an energy center. Some by-products, such as sulfur, are usually considered a waste product rather than a commodity.

On the other hand, its use as a feedstock for another industry should be considered a most efficient form of waste disposal. The products of coal gasification itself can be considered a source of raw materials for the manufacture of other synthetics such as alcohols, ketones, waxes and other hydrocarbons, and not just a source of energy for power and heat.

The emphasis of the Michigan plan is the collocation of coal conversion facilities with electric power generation or industrial activities to achieve energy conservation through the integration of processes. This energy conservation could be developed from the delivery of process steam to the coal gasification operation or the transfer of sensible heat from one process to another under conditions that are favorable for both.

For analysis purposes, the Michigan energy center team chose a coal gasification plant producing 250 million ft^3/day of gas at 300 Btu/ft^3. This is designated the reference plant in the following discussion.

Water requirements for the Michigan reference plant include both process water and cooling water. Although these requirements depend on the specifics of plant design, it was assumed that every effort would be made to conserve water through substitution of air cooling for evaporative cooling and by efforts to treat and recycle the water. Water use is defined as water not returned to the hydrologic system, because of evaporation, physical or chemical incorporation in processing, or return in an irretrievable form. The Michigan team estimated a total water demand of about 4.5 mgd.

The potential for air pollution is a major concern for any facility utilizing coal in some form or other. Among the possible pollutants of coal refineries are: sulfur dioxide, hydrogen sulfide, ammonia, carbon dioxide and other miscellaneous odors. The pollutant of major concern is sulfur, which will appear primarily as sulfur dioxide, hydrogen sulfide in the stack gases during the gasification process or when the low-Btu gas is burned. The use of the more abundant high-sulfur Eastern coals will require effective elimination of the sulfur. Coals east of the Mississippi contain 3-4% sulfur which, considering the amount of coal consumed (over 4,000 tons per day), will mandate the removal of most of this sulfur before dispersal to the atmosphere. Two methods of sulfur removal are possible; removal from the stack gases after combustion or removal in the gasification process before the gas utilization. Much early research on sulfur oxide control centered on the removal of sulfur from the flue gases rising from coal combustion. So far these "stack-scrubbing" processes have been too unreliable for sustained performance and more costly than anticipated. However, if the coal is first gasified in a process producing low-Btu gas, the resulting hydrogen sulfide can be removed before the gas is used elsewhere. This alternative seems reasonable in the tests performed so far. The main advantage of removing sulfur at this stage of the process is that only half the volume of gas needs to be treated (remembering that combustion is a process combining air with gas and increasing its volume). It also turns out that the sulfur by-product is less environmentally troublesome and more easily used for other purposes. Offsetting these advantages somewhat is the possible loss in process efficiency because of the need to cool the hot gas before it enters the H_2S removal process. Future research into methods of high-temperature sulfur removal may alleviate this problem as, for example, by the use of fluidized bed boilers.

The New Source Performance Standard for substitute natural gas plants is 1.2 lb of SO_2/MBtu. This very small allowable atmospheric discharge of sulfur may require a combination of coal pretreatment, hydrogen sulfide removal from the product gas, and the use of stack scrubbers. Data from the Water Resource Center,[19] based on calculations made for a HYGAS plant using coal with 4.42% sulfur content, indicates that only about 10% of the sulfur would eventually enter the atmosphere. Natural gas substitution plants will still require very efficient methods for the removal of particulate emissions: the New Source Performance Standard is only 0.1 lb/MBtu.

Solid waste disposal is also a concern for the coal gasification facility. The greatest volume of these solid wastes result from ash slag sludges and mining wastes. Estimates show a total amount of solid waste to be disposed of at the reference plant to be 5,000 tons per day. This waste, if used for landfill, would cover approximately 1,250 ac to a depth of 10 ft over a 20-yr period.

Energy that cannot be converted into a useful gas product because the process is not 100% efficient must also be considered a pollutant. For the Michigan reference plant the heat dissipated was assumed to be 25% of the heat content of the coal input or about 25,000 x 10^6 Btu per day.

The manpower requirement for the reference coal gasification plant estimated by the Michigan team included 164 operating personnel, 120 maintainence personnel and 86 supervisory, technical and clerical personnel: a total of 370 persons. This estimate was made to determine the possible social impact of constructing and operating such a reference plant.

An estimate of the capital investment needed for the reference plant was difficult to determine, both because of the uncertainties of cost escalations and the difficulty of specifying the exact plant technology and chosen process. Estimates range from about $100 to $300 million (in 1975 dollars) for 300-Btu gas production of 250 million cubic foot of gas per day.

Typical characteristics of a low-Btu gasification plant assumed for the Michigan energy center study are listed in Table IX.

Table IX. Typical Characteristics of a Coal Gasification Plant

Capacity	250 million ft^3/day 300 Btu gas
Total Btu Output	75,000 x 10^6 Btu/day
Heat Dissipation	25,000 x 10^6 Btu/day
Coal Requirements	4,200 ton/day at 12,000 Btu/lb
SO_2 Output Emissions	67,000 lb/day
Water Requirements	4.5 mgd
Steam Requirements	75,000 lb/hr
Electric Power	20 MWe
Plant Area	1,250 ac (including waste disposal)
Investment	$100 million (min.)
Labor Force	370

PULP AND PAPER MANUFACTURE

Several industries illustrate the advantages of collocation with energy-producing facilities, *e.g.,* the pulp and paper industry. The pulp and paper industry should be attracted to an energy center site by the assured availability of electric power and low-cost process steam, and by the availability of good transportation for both raw materials and finished products. Certain chemicals used in paper manufacture, such as chlorine and sulfur (from stack scrubbers?) may well be available from other center activities.

Estimated Future Demand for Pulp and Paper

The decision to construct a pulp and paper manufacturing capacity at any energy center will depend on future estimated demand for the products as well as the expected economic advantages of collocation. Standard & Poor's *Industry Surveys*[20] provides estimates of the annual growth of the paper industry. The annual capacity available for 1974 for all grades of paper and paperboard was about 66 million short tons, while the capacity for wood pulp was about 52 million short tons. For the three-year period 1975-77, the total additions of paper and paperboard capacity were about 6 million short tons, for an annual growth rate of 3.1%; the total additions of wood pulp capacity were 4 million short tons, for an annual growth rate of 2.6%. Of the 6-million-ton increase in this three-year period, new facilities provided about half the increases while other improvements made up the remaining capacity. If we assume the same linear rate of growth of new capacity in the U.S.

over a 10-year period of energy center construction, a total installed capacity of around 2,000 tons per day capacity would constitute about 6% of the projected growth. If such a sized plant were installed at one or more proposed energy centers, this would be a realistic goal since it would represent approximately the total increase in new facilities for the several states.

Energy Conservation

The design of a new pulp and paper mill at an energy center illustrates the opportunities for energy conservation. Gyftopoulous *et al.*[21] gives comparative data on energy requirements of the industry in 1967 per ton of product and energy requirements for new installations which would take advantage of energy conservation measures. They suggest measures including reducing drying requirements at various points in the manufacturing process, using continuous rather than batch digesters, increased recovery in waste heat boilers using bark and spent-pulp liquor, and optimum design of a combined system for generating both electricity and process steam. Many of these conservation methods, especially the latter, fit well with the energy center concept. Total savings at a new 2,000-ton/day operation based on the improved processes mentioned could be as much as $30,000 \times 10^6$ Btu per day. Nearly half this savings accrues from efficient cogeneration of steam and electricity, such as achievable at an energy center.

Pulp and Paper Mill Requirements

Table X summarizes requirements for a 2,000-ton/day pulp and paper operation. The water requirements have been reduced by 21% from a stand-alone mill since this is the approximate amount required for electric power generation. It is assumed that the power will come from the generating facility.

Table X. Pulp and Paper Mill Data

	Pulp and Paper Mill	Paper Mill Only
Capacity (ton/day)	2,000	2,000
Electric Power (MWe)	70	7.2
Process Steam (lb/hr at 60 and 175 psig)	1,260,000	121,000
Consumptive Water Requirements	9,000,000	300,000
Heat Dissipation (10^6 Btu/day)	41,000	5,800
Land Requirements (ac)	700	650
Operating Employment	4,150	3,750

While an energy center would seem attractive to any industry requiring large quantities of process steam and electric power, input and output needs must also be provided. Clearly, pulp and paper mills must be established near the source of raw materials and abundant water supplies, otherwise the economic advantages of collocation would be lost in transportation costs. Only a few locations proposed thus far would meet all requirements. The Michigan proposed plans seem most accommodating to pulp and paper mills because of the abundance of water from the Great Lakes and the proximity of usable forest timber. The Hanford site may also be suitable but the others, such as Camp Gruber, the Wasatch Front Area and Puerto Rico, do not propose to include such an industry near their center projects. These site suitability contrasts again indicate the need for careful adjustment or "tuning" of the energy center concept to the particular site.

PETROLEUM REFINING

Location Factors

Most of the energy center studies considered the inclusion of a petroleum refinery. Although the electric power requirements are not substantially higher than other industries, the potential for process steam use requires a location very near the source of steam generation and makeup water. Other factors considered important in determining the location of petroleum refineries are site availability, fuels, transportation and distribution facilities, labor supply, applicable laws and regulations, and the state and local tax structure.[22] It is usually desirable to locate a refinery near the source of crude oil, if possible, or near the normal transportation routes for crude oil. Both the Utah and Michigan centers suggest oil refining because of the proximity of crude or crude oil trade lanes and markets. Camp Gruber and Puerto Rico would be less feasible sites for oil refining because of increased crude transportation costs; Hanford and Pennsylvania excluded the possibility on several grounds which included water availability.

Refinery Data

A base-case petroleum refinery with a capacity of 250,000 bbl/day is described here as a typical size suitable for most energy centers. The data are summarized in Table XI.

Current energy requirements at a petroleum refinery result in fuel use of around 4.6×10^6 Btu per ton of refined product. Petroleum refining

Table XI. Petroleum Refinery Data

Capacity	250,000 bbl/day
Electric Power Needs	42 MWe
Process Steam Requirements	2.0 million lb/hr
Consumptive Water Requirements	12.8 mgd
SO_2 Output	59,000 lb/day
Heat Dissipation	$160,000 \times 10^6$ Btu/day
Land Requirements	1,500 ac
Construction Employment	3,000
Operating Employment	600
Capital Investment	$250 million

offers the opportunity for energy conservation through the generation of both electricity and process steam at the center. Twenty-six percent of the fuel use of a typical refinery is for process steam.[21] The 2.0-million lb/hr steam requirement would be at 125-150 psig for a 250,000-bbl/day operation. Electrical requirements of the refinery are about 4 kWh/bbl, so that a 250,000-bbl/day operation would give an electrical load of about 42 MWe.

"A modern petroleum refinery requires a large area. Most of this land is needed for tank farms to store crude oil and products. Relatively little land by comparison is used for the process facilities. An increasing amount of land will be needed for pollution control, *e.g.,* settling ponds, water treatment plants, disposal sites for oil sludge, etc. In addition, there is usually a buffer zone at least 500 feet wide surrounding the entire refinery."[23]

"It is very difficult to generalize on the land requirements for a new grass roots refinery because each site represents a special situation. A very crude estimate of minimum requirements would be 2 ac/1,000 bbl/ day for the refinery itself. Actual total land for new grass roots refineries averages about 21.8 ac/1,000 bbl/day. This generally allows room for expansion."[23] The Project Independence Report[24] estimates a much lower land requirement of 1,200 to 1,800 ac, or 5-7.5 ac/1,000 bbl/day capacity.

Petroleum Refining at the Michigan Site

Location of a petroleum refinery at a Michigan energy center would be favorable because of land availability, fuel supply, water, and transportation facilities. It is usually desirable to locate a refinery near a source of crude oil, if possible. Although some oil is produced in Michigan (about 60,000 bbl/day), this amount in itself would not be sufficient to support a local refinery of the size described here. Also,

Canada will be declining as a source of crude oil over the next 10 years. It appears that much of the future domestic oil will be obtained from offshore locations. This suggests that the Atlantic coastline would be the logical selection for new refinery locations. However, recent attempts to obtain these sites have run into serious difficulties because of the lack of suitable land and because of environmental considerations. The availability of a ready-made site at the Michigan energy center could, therefore, overcome this disadvantage and tilt the decision toward a Michigan location.

Transportation has already been mentioned as a highly important factor in locating refineries. For the Michigan site (assuming the existance of suitable harbor facilities), transportation by boat would be feasible. Crude oil from the Atlantic coast, Latin America or the Middle East could come through the St. Lawrence Seaway. Transportation of the finished products could be also handled by boat or by pipelines. In addition, it should be recognized that a sizable portion of the product would be required for state and energy center consumption.

A major expansion of domestic refinery capacity is considered not only likely but necessary. The newer units constructed will offer the capability of processing high-sulfur foreign crudes. Using the rough estimate of 10 million barrels daily of new U.S. refinery capacity during the construction of the proposed energy centers, a 250,000-bbl/day refinery would represent around 2.5% of the total additional U.S. capacity. Thus this sized petroleum refinery would be a reasonable adjunct to any energy centers.

The possibility of constructing energy centers in Michigan on or near the Great Lakes provides a favorable location factor in the availability of water. Most refinery operations, from primary distillation through final treatment, require very large volumes of process and cooling waters. Estimates of water usage at petroleum refineries indicate a water intake of nearly 400 gal/bbl of crude oil and water discharge of 340 gal/bbl. Excluding water used for steam generation and boiler feedwater, which will not be required in the collocated refinery, the water consumption for a 250,000-bbl/day plant becomes 12.8 mgd.

Collocating Refineries at Energy Centers

Two characteristics of petroleum refineries—the need for large quantities of process steam and the need for nearby markets—make them particularly attractive for inclusion in some energy centers. A major reason for large-scale collocation of energy facilities at a center was the ability to supply large quantities of process steam economically. A refinery located at an energy center stands a good chance of having a ready nearby market for many of its products and, therefore, lessened transportation needs, resulting in lower distribution costs and the expenditure of less transportation energy. It should be emphasized that this is another example of how collocation and economies-of-scale at a well-coordinated energy center can achieve considerable energy conservation. Less clear, but possibly also very important, is whether certain by-products and waste materials can be effectively utilized among other elements of the energy center. For example, can the sulfur extracted from high-sulfur oil be economically used in other processes, *e.g.,* pulp and paper manufacture, in the form of sulfuric acid? It should be remembered that the production costs need not include the cost of extraction from fuel as this must be done. There is a saving, too, in not having to dispose of the waste materials.

CHEMICAL INDUSTRIES

Chemical industries are expected to be an important element of any energy center because of the availability of process steam and electric power. However, it is very important to match the production facilities with the easily available raw materials and accessible markets. Several studies have previously attempted to find a good match among all factors starting with the Puerto Rico study in 1970 and closely followed by the Utah study by the Dow Chemical Company. The Dow study is a good example of an attempt to exploit the particular resources of a study area.

The Wasatch Front is the most densely populated area in Utah. Comprised of Davis, Salt Lake, Utah and Weber counties, it contains 78% of the state's population, 80% of the commerce and 81% of the industry. In recent years, this area has grown at double the average rate for the U.S. Among the notable resources of the area is that 40% of the state's manageable water supply flows into the Wasatch Front. Although this water has become brackish by the time it nears the Great Salt Lake, this has been viewed as a resource to be exploited. The Dow study is a preliminary look at the feasibility of constructing a 500- to 1,000-MWe power plant in conjunction with an industrial segment in the Wasatch Front area. Emphasis was placed on utilizing those natural resources that were available to the best advantage. Thus, desalting plants would be necessary to provide boiler make-up water and process water, but the extracted salts would then be the source of sodium, chlorine, bromine, etc. Crude oil and coal is abundant in Utah and was considered in determining the nature of the industrial complex chosen for the Dow study. The Dow study was primarily an economic and marketing study which examined the feasibility of a certain industrial configuration at the time of the study. It therefore excluded the use of some raw materials such as coal and potash because the market outlook at that time was not favorable. Circumstances have already changed considerably since the study was undertaken so the industrial configuration recommended by Dow may no longer be the best solution. The results do, however, provide a good example of how an overall design utilizing available resources can be combined into an industrial-water-power complex. The Dow industrial complex block diagram was shown in Figure 14. Note in particular how the input materials of brine, water and crude oil result in the production of many useful, salable products such as magnesium, vinyl chloride, polyethylene and fuel oils. Another important consideration is to note how many intermediate products exist which provide the raw materials for the end product processes. These include hydrogen and chlorine propylene caustic. Some intermediate products are also available for sale as well as for further processing. This allows for further flexibility in "fine tuning" the entire process as more intermediate products can be diverted for direct sale whenever market conditions are favorable and, conversely, the intermediate products can be converted to other end products. This flexibility results from the appropriate design of any large, complex industrial center and is a particular feature of energy centers that should be exploited.

A summary of the requirements and outputs of the industrial center evaluated for the Wasatch Front area are given in Table XII.

**Table XII. Characteristics and Requirements
of Wasatch Industrial Complex**

Product Capacity, All Products	2,900 million lb/yr
Magnesium	50
HCl	18
Methyl Chloroform	40
Carbon Tetrachloride	20
Perchloroethylene	60
Vinyl Chloride	500
Polyethylene	450
Propylene	320
Fuel Oils	840
Caustic	320
Propylene Oxide	200
Brine Requirements	18,600 mgd
Water Requirements	3,420
Crude Oil	4,270
Electric Power	310 MWe
Process Steam	2,100,000 lb/hr
Land Requirements	5,000 ac
Employment	1,000 persons, approximate
Startup Capital	$418 million (est. 1978)

The Dow study concludes that an industrial chemical complex could be constructed near Salt Lake City which would utilize Utah's raw materials, provide additional employment, furnish an industrial base for further development, and yield adequate profits.[5] This chemical complex offers an excellent example of balancing the area's needs, market potential and available raw materials, and interlocking all units to achieve energy and material conservation.

The Michigan Chemical Center Concept

The previous section described the Dow chemical complex at the Wasatch Front area as a good example of matching requirements to available resources and markets. This section describes another chemical configuration, one suitable for the Michigan site, in order to emphasize the contrasts that result when different geological areas are selected.

Chemical industries are expected to be an important component of the Michigan energy center. The specific chemical products to be produced at the center will depend on the availability of raw materials, markets for the products, and the economics of producing them at the site. Without precisely defining a group of chemical products for the Michigan site, the study team did outline the major characteristics of the chemical industry complex. A chemical complex in Michigan could be based on a combination of raw materials characteristic of the Great Lakes area, including coal, petroleum, timber, salt, limestone or metal ores. Raw materials might also be derived from other industries at the center. For example, coal liquefaction could be a good source for aromatics; sulfur might be available in large quantities from the desulfurization of coal.

The chemical industry has grown at nearly twice the rate of other industry. Although there are indications that this trend may not continue, it is the general opinion of several financial analysts that the chemical demand will increase at a rate of 2-3% more than the average expansion rate of the total economy. An estimate of the magnitude of the proposed Michigan chemical complex can be reached by using these predicted growth rates. Based on current sales and this projected growth, the increased U.S. annual capacity of low-density polyethylene would amount to around 7,000 million pounds. Thus a plant capacity of 300 million pounds, the capacity assumed for the Michigan energy study, would meet about 4.3% of the nation's requirements. This would seem to be a reasonable-sized plant for that part of the country which is, incidently, about the capacity of the Wasatch Front area.

Although specific chemical production facilities were not determined for the Michigan study, it was expected that the configuration would be similar to the Dow study and, therefore, many characteristics of the chemical complex could be estimated. The capital investment would be in the neighborhood of $500 million and would provide employment for about 1,000 persons.

The electric power requirements would be approximately 300 MWe and the complex would use over 2 million pounds of steam per hour, and 6 million gallons per day of process water, and 20 million gallons per day of lower-quality cooling water. These requirements make it easy to see why locating such a chemical complex near a Michigan energy center would be advantageous. First, there is the abundance of water from the Great Lakes. Second, the energy-generating facilities would provide the electric power and the tremendous quantities of process steam. And third, it is expected that the other center industries would be sources of processing materials as well as ready markets for the chemical outputs. Table XIII summarizes the chemical complex characteristics for the Michigan center.

Table XIII. Chemical Complex Data

Capital Investment	$500 million
Annual Sales	$250 million
Electric Power	300 MWe
Process Steam Needs	2 million lb/hr
Consumptive Water Requirements	26 mgd
Heat Dissipation	$93,000 \times 10^6$ Btu/day
SO_2 Production	48,000 lb/day
Land Requirements	500 ac
Operating Employment	1,000

Chemical Complexes at Other Energy Centers

Most energy center studies considered the inclusion of a chemical manufacturing complex within the scope of the study. This acknowledges the advantages of nearby electric power and process steam. In addition, most studies recognized the desirability of matching particular chemical processes to available markets and raw materials. Very few studies went into specific detail on a chemical industry configuration This was, in part, because such studies are vastly more complex than other studies so far commissioned and, furthermore, such details do not affect the overall feasibility of energy centers to any great degree. The disadvantage to this lack of depth is the paucity of information on the benefits of carefully matching the output-input requirements of the chemical indus tries to other energy center units. At this time we are unable to evaluate the economics of adjusting the elements of the energy center so as to make the waste products of some units the raw materials of other units.

MINI-STEEL MILLS

The U.S. steel industry needs about 25 million tons of added capacity before 1980 and should replace perhaps an equal amount. However, the major U.S. steel producers are not gearing up to meet this demand of high investment costs, uncertainties regarding future government policy and price uncertainties resulting from the impact of imported steel. An alternative to the construction of a full-scale steel mill using iron ore as its raw material is the so-called mini-mill, with an investment requirement only one-fourth that of a large integrated mill.

The idea behind the mini-mill that allows it to compete with larger integrated mills is proximity to customers and raw materials, and a concentration on basic lightweight products. The bulk of production would be carbon steel, the lowest grade. Typical mini-mills use scrap steel and scrap iron as their primary raw materials. They melt the scrap in electric arc furnaces and cast and/or roll the resulting steel to produce a limited range of commonly used products. The specific products produced depends on the location of the steel mill and its relation to the markets. Usually, the products are at the lower end of the size, quality and price scale, *e.g.,* concrete reinforcing bar, hot-rolled bars, wire rods and light shapes.

Annual production capacity of mini-mills runs from 100,000 to 300,000 tons and is recently increasing even more to achieve economies of scale. The increased production is accomplished by broadening the product line and including items of increased complexity and higher quality.

Many mini-mills use electric arc furnaces. Thus, a fundamental requirement for successful operation is a guaranteed supply of electricity at a competitive cost. Also, most small mills require an assured supply of natural gas or other fuel to reheat castings before forming them in the rolling mill.

Transportation costs for both raw materials and finished goods are a major expense in the steel industry and so take on special importance for the mini-mill. Competitive rail and truck transportation between raw material sources, the mill and its customers is a necessity. Because mini-mills use scrap, they are outstandingly efficient and remarkably non-polluting. The EPA has estimated that steel made from scrap uses 74% less energy than that made from iron ore. Mini-mills have very desirable environmental characteristics; the small amount of fumes and dust generated by electric furnaces is easily contained and filtered. Water is not chemically contaminated, but used and reused for cooling. The EPA also estimates that mini-mills create 84% less pollution than steel mills using iron ore. Furthermore, these mills dispose of otherwise troublesome junk.

Mini-Mill Data

A typical mini-steel mill would include scrap-handling facilities and provide an annual melting capacity of 150,000 tons. This would include two 30-ton per melt electric furnaces, a continuous casting machine, billet reheating furnaces and limited rolling mill equipment capable of producing simple shapes in small sizes.

Table XIV summarizes some of the features of a mini-mill suitable for inclusion at an energy center.

Table XIV. Mini-Steel Mill Data

Capacity	150,000 ton/yr
Electric Power	15 MWe
Heat Dissipation	$3,000 \times 10^6$ Btu/day
Land Requirements	100 ac
Capital Investment	$20 million (1974 dollars)
Operating Employment	500

The Mini-Mill at an Energy Center

The mini-mill has several features that make it an excellent candidate for inclusion in some energy centers. The Michigan energy center study seriously considered the mini-mill. The location of mini-mills near the consumer gives them a competitive edge over more distant producers because of lower transportation costs. Mini-mills must be near their markets in order to profit from small orders and to offer fast delivery. The maximum economic limit of a mini-mill territory is about 300 miles.

Michigan's automotive industry makes the mini-mill concept very attractive not only because of the ready markets but also because of the large quantities of scrap steel available. Over 30% of the total U.S. production and import of steel products producible by mini-mills are used within a 300-mile radius of the southern half of Michigan's lower peninsula. In 1972, this Michigan market totalled annual sales of 5.7 million tons of concrete reinforcing bars, hot-rolled bars, light shapes and wire rods. Of this total, only 400,000 tons are presently produced by mini-mills, indicating a tremendous market potential. The products of local mini-mills might penetrate this market by nearly 3 million tons by 1985. The existing mini-mills would be too distant to be a leading competitive factor in this 5.7-million-ton market.

The Office of Economic Expansion of the Michigan Department of Commerce has been actively promoting the construction of mini-mills in Michigan. At one time the North Star Steel Company was planning to locate a $50 million mill near Muskegon which would have had a capacity of 300,000 tons per year and employed 500 persons. However, because of local environmental objections, the company has apparently abandoned the site and is looking for another. If Michigan decides to construct an energy center, the inclusion of a mini-steel mill seems obvious.

AGRICULTURAL AND AQUACULTURAL SYSTEMS

It may be restated that the primary objective of an energy center is to utilize, as well as possible, the heat content of a fuel while minimizing the environmental impact of the process. Most previous discussion in this book has been directed toward improving the overall efficiency of fuel utilization. Yet despite all efforts at conservation, it is still impossible to achieve more than 50% efficiency in fuel utilization. One reason is our insistent use of electricity. It is, of course, a most attractive form of energy—it is easy to transport, and converts to work or heat very efficiently and is very clean. However, frustrated by the laws of thermodynamics, no more than one-third of the heat of combustion can be converted to electric energy. The remaining heat content appears as waste heat. More disappointing is that this is low-quality heat; that is, it is available only at low temperature. In order to achieve maximum efficiency of electric power generation at the turbine, it is necessary to reject the waste heat at temperatures near, or somewhat above, ambient. This 80°F heat can no longer drive machines or heat buildings or be used as process steam. Unless uses can be found for low-quality heat, this energy is not only lost, but may also cause detrimental effects on the environment. This section discusses one of the more promising uses of this "waste" heat. Plants and animals require an environment whose temperatures lie in the range of the rejected heat from power plants. In a way, then, living things might be able to use such rejected heat to utilize their primary energy or food source more efficiently. This possibility leads to the use of power plant waste heat for various agricultural or aquacultural systems.

New fossil fuel electric generation facilities can achieve an overall efficiency of around 40% while nuclear plants can reach over 30%. Thus for our basic energy of 20,000 MWe, we can expect from 80×10^9 to 130×10^9 Btu per hour to be rejected to the surrounding environment.

Such concentrations of waste heat discharges can significantly alter natural biological patterns in the environment. For example Bryan[24] has tabulated data that suggest that a temperature increase of only one or two degrees may be harmful to the ecosystems of certain lakes and inland streams. On the other hand, a temperature increase by waste heat can be beneficial for the growth of some crops and fish cultures.[25]

Using conventional heat dissipation methods, the construction of a large-scale energy center in all but the most remote areas may be in conflict with increased concern for environmental protection. Hence the possibility of alternate methods of dissipation using agricultural or aquacultural systems has the advantage of reducing the environmental impact of power generation and increasing the return on the investment for the utilities.

Before describing some of the possible systems recommended for this purpose, several problems should be mentioned. The proposed systems will try to avoid these problem areas. First, single agricultural or aquacultural uses generally have seasonal characteristics that make them unsuitable for continuous heat utilization. Therefore, attention should be given to multiuse integrated systems in an attempt to achieve year-round benefits. Second, the waste heat is usually in the form of large quantities of warm water. Either very efficient low-temperature heat exchangers are necessary or the water must be utilized directly. The latter is difficult since, for example, using the water directly for soil warming could also cause flooding.

The discharge of waste heat from steam power plants represents a considerable amount of energy which may be recovered under certain favorable circumstances. The most important parameter of any heat source is its temperature. Low-grade or low-quality heat is characterized by low temperature. High-quality heat, on the other hand, is represented by high temperatures. The boundary between the two can be placed arbitrarily at the temperature of boiling water at atmospheric pressure. This temperature is roughly the point at which a fluid can or cannot provide useful work. This distinction is important for an assessment of the potential benefit of waste heat from power plants. Thus high-grade heat is available for process steam, performing work or district heating. Low-grade heat has limited applications. Unfortunately, one half to two thirds of the heat content of the primary fuel of power plants is discharged as low-grade heat. Any efforts to reduce this percentage, by producing more process steam, for example, always means less electricity will be generated.

The low-grade heat discharged by power plants is in the form of large quantities of warm water 30°C or perhaps 20°C above ambient. The only promising applications suggested so far are certain biological applications.

These include the areas of fish- and crop-growth enhancement, soil warming, warm-water irrigation and greenhouse warming. Some of these potential applications will be discussed in the following sections.

Fish Farming

It has been estimated that about 95% of the catfish sold in the U.S. come from aquacultural enterprises—fish farms ranging from small ponds to elaborate fish culture complexes. Similarly, nearly half the trout available to consumers comes from huge trout farms in the south and west. Lately it has become apparent that this type of aquaculture may also be feasible in the northern states in ponds continually fed with the warm wastewater from power plants.

It has been known for some time that raising the temperature of a body of water above its ambient temperature can result in a change in the species composition of native aquatic organisms. Such an effect can be viewed as detrimental or beneficial depending on which species are considered desirable for harvest. Some fish, such as carp and catfish, have a high heat tolerance while trout and salmon have a low heat tolerance. Saying some fish have a high heat tolerance may be misleading since, in fact, these fish have higher growth rates and actually prefer water temperatures of from 27 to 34°C. This temperature is just about the same as the warm water discharge from steam power plants.

Although some fish thrive at these elevated temperatures, it is necessary that their food-chain organisms exist at these elevated temperatures also, unless the fish are to be fed externally. For the most part, it has been shown[26] that fish are frequently more sensitive to elevated temperatures than most food-chain organisms. This implies that sufficient food organisms will be present, though they may be of a different kind, to support the harvested crop. Therefore, it is necessary to sustain the food-chain organisms as well as to provide appropriate conditions for spawning and growth of the fish crop. Of course, under some conditions it may still be economically feasible to artificially supply fish feed and undertake the spawning and early growth at other locations, but this would no doubt reduce the possible return on investment.

At the present time, carp is not considered a preferred food source in this country; however, catfish, another high-heat-tolerant fish, has a ready market. Cultured channel catfish exhibit many characteristics that make them a prime candidate for use in a warmed-water environment. These characteristics include:[27]

   1. Channel catfish are nutritious, high in vitamins, minerals and protein.
      Properly managed farm-raised catfish have excellent flavor. They

    contain little or no saturated fat and are low in carbohydrates. They can be processed and utilized in a variety of ways.

2. Stability in the supply and quality of intensively cultured catfish make them more adaptive to marketing than the variable commercial catch of wild fish.
3. Current and future demand for catfish appear strong. Recent estimated catfish consumption growth is 7% annually.
4. The sport-fishing aspect of catfish from fish-out ponds may become even more popular.
5. Vertically integrated techniques of mass production, such as used by the broiler industry, can be applied to catfish.
6. Catfish have a very high feed-conversion ratios compared to other farm products. The catfish feed-to-meat conversion ratio is around 1.5:1. Broilers convert at about 2.3:1, hogs at about 3.5:1 and beef at 8.0:1.
7. It is expected that catfish farming would cause new investment and employment in the commercial producing sector and related businesses.
8. The preferred water temperature for catfish growth is 18-32°C, approximately the temperature of discharge water from most power plants. Such temperatures can be maintained in fish ponds nearly year round, greatly extending the growing season.

Although catfish are native to almost all waters in the U.S., nearly all farm-pond production and intensive cultivation has been concentrated in the river basins of the Gulf Coast on the southern half of the Mississippi River drainage system. The production of channel catfish has become specific to this region because of superior climatic conditions, abundant water supply and impervious soils. Catfish are also considered a premium food fish by the people of these regions, and so a well-developed processing and marketing structure exists in this area.[27]

Pond culture of catfish is the predominant means of production, but the use of raceways is growing in use and importance. The major characteristics of pond culture are: (1) lowest capital requirement per acre, (2) waste removal is more of a problem than with other methods, (3) harvesting methods do not insure 100% recovery, (4) if disease is contracted, the costs will be proportional to the size of the pond and (5) available dissolved oxygen is more difficult to control.

Raceway or running water culture, either wholly or partially, has the following characteristics: (1) more intense cultivation per acre is possible, (2) harvesting costs are lower than for ponds, (3) capital construction costs are greater than pond construction, (4) dissolved oxygen depletion is made up by the freshwater brought into the system and (5) financial loss due to disease may be lower than that for ponds.

Although it is not yet possible to determine whether fish culture at an energy center will be a profitable undertaking, several successful ventures have already taken place at single power-generating facilities. Tilton and Kelley have been raising channel catfish in cages in the discharge canal of the Texas Electric Service Company at Lake Colorado City, Texas.[28] They found that temperatures in the canal from January 10 through January 26, 1970, were high enough for channel catfish to have a 45% weight increase; fish kept in an unheated pond failed to gain weight. Based on these results, farming of channel catfish in a power plant discharge could be extremely profitable at current market prices.

Costs and Rates of Return for Pond Culture

Michigan State University, for example, presented estimated costs and rates of return for a farm consisting of eight 20-ac ponds. This study was mostly a paper analysis but was felt to fairly reflect the possibility obtained using the most advanced technology, production practices and management. Other assumptions included the availability of suitable land, a 210-day growing season and other reasonable criteria for making a sensitivity study. Costs were based on stocking the ponds with 6-in. channel catfish at the rate of 2,000 per surface acre of pond. For stocking done on March 15, harvesting would occur on October 15; the average weight per fish was assumed to be 1.25 lb. The cost calculations were also based on an assumed mortality rate of 5%, a 1.6:1 feed-conversion ratio and a per-acre production of 2,357 pounds of fish.

Pond characteristics must be included in the assumptions of the study since they effect the cost calculations. In this study, the pond characteristics included a 14-ft crown, a 3:1 side-slope ratio, a 50-ft base, a minimum depth of 4-ft with a 2-ft freeboard. Water quality, temperature and temperature stability are important factors that are difficult to define, but it was assumed that the minimum conditions would be met by the typical discharge from power plants. The fish feed would be in the form of pellets used at the rate of 3,610 pounds per surface acre of water. Harvesting would be by the "mechanical haul seine technique."

Assumptions concerning investment costs were based on both a total and per-surface-acre-of-water schedule for equipment and facility requirements. The initial investments on a unit production basis tend to decline as the size and number of ponds in a facility increase. However, there are decreasing returns with increasing scale in harvesting, increasing financial losses due to disease and increasing costs for waste removal in larger ponds. The initial investment requirements are shown in Table XV.

Annual cost includes ownership costs, depreciation, interest on investment and the annual operating costs. Typical annual costs based on the assumptions made in the study is given in Table XVI.

**Table XV. Estimated Initial Investment Requirements
of Pond-Cultured Catfish**

Item	Cost
Land	$ 64,000
Pond Construction	40,000
Water Wells and Supply Pipe	40,242
Feeder and Feed Storage	1,950
Disease, Parasite and Weed Control	1,200
Harvesting Equipment	8,652
Miscellaneous	17,226
Total Investment	173,270
Investment per Surface Acre of Water	1,216

**Table XVI. Estimated Annual Costs of Producing
Catfish for Food by Pond Culture**

Item	Dollars	%
Annual Ownership Costs	$ 7,784	8.48
Depreciation	7,984	8.69
Interest of Investment	3,145	
Subtotal	18,913	
Annual Operating Costs		
Repairs and Maintenance	3,677	4.23
Fuel, Feed, Fingerlings, etc.	46,706	50.89
Labor	19,575	21.33
Subtotal	69,958	
Total Costs	88,871	
Total Production (lb)	327,574	
Cost/lb	0.2713	

This table shows that the production of catfish by pond culture might
cost around 27¢/lb. If the market price received for the farm-produced
fish is 34¢/lb, for a fish farm of this size there would be $23,221
available to cover annual ownership costs, depreciation, interest on investment and operating costs of the total system. This would provide for a
13% rate of return on investment, which seems reasonable. Similar studies
have been performed on other types of fish culture including raceways
and semi-raceway cultures, and trout farming. In all cases, the results
are favorable for fish culture, depending, of course, on the market climate
at the point of sale. It should be noted that these economic studies were
viewed as independent entities separate from other resources or production
capabilities. Such fish cultures, then, would no doubt be even more
attractive when located near energy centers, enabling them to take

advantage of cost-reducing complementary relationships among other units and of the low-cost warm water.

The cost analysis in the preceding paragraphs portrays the advantages to fish-culture systems located near an energy center. There are also advantages to the energy center, most notably a means of utilizing the immense quantities of waste heat. These fish-culture ponds provide an excellent example of the symbiotic relationships which benefit both elements of a system. Here the waste heat provides the thermal environment for optimal growth of catfish and the catfish pond provides dissipation of heat through convective and evaporative heat transfer at the water surface. Heat dissipated by the ponds will vary throughout the year with a maximum heat dissipation of 2.21×10^7 Btu/hr-ac during the winter months and a minimum value of 6.14×10^6 Btu/hr-ac during the summer months. Water lost due to evaporation is approximately 0.39 in./day or 10,589 gal/ac-day. The main advantage of fish culture near energy centers is not the direct use of energy, but rather the use of the already-needed cooling ponds.

Where the energy center is to be located on an ocean, such as in Puerto Rico, there is the possibility of saltwater production (mariculture). Some species, like oysters, are able to live successfully at temperatures above and below those required for spawning. There are problems of providing synchrony for the oyster larvae and the necessary food species. The Long Island Lighting Company at Northport, NY,[29] experimented with the mariculture of oysters in the heated discharge from the nuclear power plant. Mihursky[30] has proposed the use of oysters as filter-feeding organisms in a complete recycling of organic matter, utilization of residual heat for waste treatment and district heating. The results were enhanced growth of algae, zooplankton and oysters and clams. It has been noted that the larvae of lobster do not tolerate temperatures below 15°C and may have increased growth rates as well. The high food value of lobster is an effective incentive for mariculture using the waste heat discharges from energy centers.

Grain Drying

Over 50% of the total energy inputs to corn production in some areas is used for drying. Many drying facilities currently use fuels in short supply—propane, natural gas or electric heat. Clean fuels are required to keep the grain free of the contaminants of combustion. As an alternative to these expensive fuels, it has been suggested that waste heat from energy centers be used both as a source of inexpensive energy and as a means of reducing the effects of waste heat. Several studies and field trials have suggested the use of a solar-type dryer using a warm water-to-air heat exchanger. The design used in such drying systems is a fixed-bed bin with a maximum grain depth of 12 ft and a back-up electric heating system. The waste heat required for the one-month drying period is about 80,000 Btu. In many areas proposed for an energy park, a grain-drying facility with a 10-million-bushel capacity would be feasible.

Greenhouse Heating

The concept of using waste heat from power plants for heating conventional greenhouses in winter and for evaporative cooling in summer was initiated at Oak Ridge, Tennessee. Many growers are interested in utilizing waste heat as a replacement for natural gas and other fuels subject to uncertain supply and rising costs. Natural gas, furthermore, is being allocated to industries more critical to the national economy than greenhouse heating. A recent Michigan study[31] has shown that such a concept may be applicable for Michigan climatic conditions, using a greenhouse system with a plastic roof and a fin heat exchanger.

A variety of vegetables and flowers can be grown in greenhouses heated by waste heat. These include tomatoes, leaf lettuce, cucumbers, eggplant, spring bedding plants and seasonal flowers. However, because heat losses by radiation are high, only 0.45 ac of greenhouse can be heated with waste heat per MWe of generated power. Consumptive water loss is 2% of the flow using wet-pad heat exchangers.

Irrigation and Soil Warming

The use of the warm water discharge from power-generating plants would seem to be an excellent way of providing sufficient moisture and an optimal growing temperature for many agricultural crops. Unfortunately it is difficult to provide both features at the same time because the water required to appreciably warm the soil would supply too much water for growth. Conversely, the heating effects of irrigation processes are not long-lasting. The alternatives are to provide only heating, only irrigation, a partial combination of both or to achieve both heating and irrigation by the use of two systems for one crop. In any case, either compromises or cost increases detract from the use of large quantities of waste heat in this manner.

The water quality is the critical condition for irrigation purposes.[31] Several horticultural and agricultural studies and demonstrations have shown that the benefits of thermal discharges are primarily a result of the moisture itself, not the heat energy of the water. The additional moisture in the soil can increase the yield and quality of some crops, and by using a sprinkle irrigation system, it can modify naturally occurring microclimatological conditions of temperature and humidity. This climatological modification might help prevent frost damage by use of water spray, especially in orchards.

It has also been demonstrated[31] that thermal water can be used successfully for plant cooling and protection against sunburning of fruit and nut crops. During hot summer days, with low relative humidities, the trees and plants as well develop "water stress"—wilting leaves. Activating the sprinkle irrigation system not only lowers the air temperature by 5-10°, but also increases the humidity. "Sunburn" to plants can thus be eliminated by this "microclimate."

The emergence, vegetative growth and yield of some crops are increased when the soil temperature of 30°C is maintained, which is approximately the discharge temperature of the cooling water of most power plants. Soil warming with waste heat can be achieved by placing a grid of PVC pipes in the ground with heated water flowing through the pipes. Although there are no commercial-size soil-warming operations yet, many studies of soil heat transfer and crop growth have been made. This includes an attempt to apply optimal control theory to determine the pumping policy that will produce the best soil temperature over the largest area, thus producing the most heat dissipation at the least cost.

Vegetable crops with a yield increase sufficient to use a waste heat production technique include tomato, snap beans and sweet corn. There may be marketing constraints on the acreage of these crops. Field crops such as soybeans, corn and peabeans do not have acreage constraints but the rate of return on investment from soil warming is marginal.

Irrigation can provide for heat dissipation corresponding to a rate of 100-125 acres irrigated for each MWe of generating capacity. On the other hand, the heat dissipated by a soil-warming system is only about 12 ac for each MWe, which reflects the fact that direct water application for soil warming would produce ten times more moisture than adequate for irrigation. Diverting perhaps one-tenth of the water in the soil-warming pipes to sprinkle irrigation would improve the heat-dissipation capabilities slightly. More importantly it would prevent the soil from drying out near the pipe, which would create a reduction in heat transfer.

Waste Treatment

Waste heat can be used to improve both secondary and tertiary municipal waste treatment. As with most biological processes, an appropriate and usually warm temperature speeds up the process. Economic benefits in secondary treatment come from the resultant increase in plant efficiency. Secondary treatment through the activated sludge process, reverse osmosis, carbon absorption and the ion exchange processes are all improved by heating with waste heat.

The benefits of waste heat for the conventional activated sludge process have been analyzed in detail by Agardy.[32] He described several heat transfer systems and showed that the optimal temperature for operation is 26°C. For a waste treatment plant located adjacent to a power plant, net savings are around 1.1¢/1,000 gal of sewage. This would be a net savings of $40,000 per year if the system were used to process wastes from a city of 100,000. For the system considered here, 0.11 million Btu/hr can be dissipated.

Several systems[31] for combining tertiary treatment with aquaculture are in the pilot stage. Coastal sites in Massachusetts and Yugoslavia were used to grow diatoms from wastes. Ninety percent of the diatoms were harvested when the medium was used to grow oysters. A similar system is being used to grow striped mullet in Florida. In Oregon, a culture of the green algae *Chlorella pyrenoidosa* and *Scenedesmus quadricuda* in a pond heated with a grid of pipes carrying waste heat is being studied. Analysis of heat dissipation rates of the ponds show that 1.63 ac can be maintained at an optimal temperature of 25°C for each MWe of power generation. In the winter, 0.54/ac per MWe can be maintained.

A pond culture is being tried in Ontario for growing the easily harvested and equally nutritious blue-green algae, *Spirula maxima.* However, the algae probably should be harvested by fish or shellfish since the harvesting costs are unknown and a marketing structure for the algae does not exist. Both marketing and legal factors for a conservative system of growing oysters on sewage products have been reviewed. In this system it is proposed that the tertiary treatment system be used to produce chopped carp for protein feedmeal. In this system, nutrients in the secondary waste water are removed in the growth of *Chlorella pyrenoidosa,* which is harvested by the algae-feeding fish such as *Tilapia mossambica* and silver amur. Conveniently, the optimal temperature for both fish and algae is near 30°C, similar to the discharge water of power plants. Estimates for these conditions give a conservative production of 40 metric tons of algae per hectare per year which can be converted to 8 metric tons of fish.

There are other agricultural and aquacultural applications for waste heat effluents, including some rather esoteric two- and three-stage processes such as sewage-to-algae-to-zooplankton-to-fish-to-food processing. Yet they are all based on biological processes and the amount of waste heat dissipated by any biological process is likely to be about the same and mostly limited by the available land area. It will be necessary to examine only one or two processes to determine the feasibility of utilizing waste heat from power plants. Estimates of the feasibility in turn depend on several site-specific factors, thus it will be necessary to describe some of the planned agricultural applications of several proposed energy centers.

Agriculture and Aquaculture at Camp Gruber

In the Camp Gruber study[2] it was recognized that large volumes of reject heat and water would be released from the electric power-generating installations. The large once-through cooled plants would discharge at least 1,000–1,500 cfs of heated water (15-20°F above ambient); cooling tower blowdown might be 10-50 cfs of heated water. This could provide an effluent temperature on the order of 30 to 45°C. Depending on plant design and source, it is possible that large volumes of reject heat and water will be available at Camp Gruber. The ability to use or dissipate significant quantities of heat and water in an agricultural production and processing complex will depend on the requirements for such heat by feasible agricultural production units.

In the Camp Gruber study, 34 conventional and unconventional agro-industrial production units were identified which could be candidate systems for inclusion in an integrated complex. These units were classified into five groups (see Table XVII) based on the type of production, *i.e.*, food processing or greenhouse systems or field crops irrigation.

Table XVII. Agri/Aquaculture Applications Proposed for Camp Gruber

I Foodstuff and Processing (120°F)	II Controlled Growth Systems (80-120°F)	III Livestock and Marine Production (70-80°F)	IV Low-Intensity Agriculture (60-70°F)	V Irrigated Agriculture (Below 60°F)
Vegetable Processing	Hydroponic and Greenhouses	Dairy	Frost Protection	Alfalfa and Seed Crops
Hay Processing	Vegetables	Poultry	Crop Drying	Soil Warming
Meat and Fish Processing	Seedlings	Swine-Sheep Beef	Residue Drying	Growth Enhancement
Other	Flowers	Aquaculture Systems		

It was assumed in the Camp Gruber study,[2] that each of these five functional groupings were distinguished by the range of operating temperatures as a result of the type of process involved. For example, the first group was concerned with foodstuff and feedstuff processing. This group of agro-industrial production units operates at the highest range of temperatures of all the groups, perhaps with a range of 50°C and higher (reaching the ranges of process steam).

Table XVII shows conceptually the proposed five groups of production units with decreasing temperature. It should be noted that some of the processes included in the Camp Gruber study form a rather smooth transition in temperature from the use of waste heat to the use of process steam. As use is made of higher and higher temperatures, it will be necessary to either produce less electricity or to degrade the efficiency of the generation. Of course, the overall rate of fuel use may still be improved. Nevertheless, such questions as to the trade-offs desired and their effects on the efficiency of the overall system is a very complex fine-tuning process that perhaps can only be obtained by large-scale computer simulation. At any rate, neither the Camp Gruber study nor this author has the resources available to determine the appropriate and optimal mixes of any of the agricultural processes.

The primary siting requirements for aquaculture-agriculture using waste heat are a source of waste heat and ample land.

For the Camp Gruber study, they assumed that waste heat would be used for a combination of greenhouse, normal cropland, and process industry uses. The greenhouse and process industries would be located on the nonagricultural Camp Gruber lands adjacent to the town of Braggs. The normal cropland is assumed to be in the general vicinity of Braggs or just northwest of the camp towards Fort Gibson. The water used to carry the heat for these uses would be recycled back to the energy facilities as appropriate. A specific configuration and land use requirement was not determined.

Agriculture at the Puerto Rico Energy Center[4]

Although agriculture production in Puerto Rico has been relatively constant in recent years, the demand has been increasing greatly. This rising demand has largely been met by increased imports of such commodities as citrus fruits, peppers, avocados, plantain, tomatoes and yautia (similar to the potato). Imports are projected[4] to reach a value of over $30 million by 1980. This offers a ready-made market for local production of these crops. Additional markets exist in some of the larger cities on the east coast of the U.S., where some Puerto Rican produce can be competitive, especially on a seasonal basis.

A study was made of a 500-ac commercial farm operation for those crops showing a high unit profitability. Although the Puerto Rico plans call for desalting plants at the West Aguirre energy center site, the desalted water at around 45¢/1,000 gal would be too expensive for agriculture near the Puerto Rican center. Feasibility depends upon accessibility to both export and home markets, the availability of water from the existing South Coast Irrigation District and the ready availability of power for water pumping. The cost estimates made in this study indicated that a number of crops would provide a reasonable rate of return on investment, including the following:

Crop	Acres
Orchard	
Mangos	50
Avocados	40
Grapefruit	30
Semiorchard	
Plantain	100
Root Crops	
Yautia	50
Sweet Potatoes	50
Vegetables	
Tomatoes	100
Peppers	35
Squash	15
Buildings and Roads	30
Total	500

As part of a program to implement these changes in agriculture, the Puerto Rico study team proposed that a demonstration farm be established to prove the economics of a 500-ac site in the West Aguirre Energy Center. They estimated capital costs, annual cost and income for a 1975 start as follows:

Fixed Capital	
Buildings	$150,000
Farm Machinery	152,000
Irrigation Equipment	300,000
Total less Land	602,000
Working Capital	
Accounts Receivable	75,000
Crops in Progress	75,000
Total	150,000
Operating Costs and Profit	Years 1-4
Sales	780,000
Costs	-585,000
Operating Profit	$195,000

Agricultural Applications at Other Energy Centers

Almost every energy center study undertaken to date included at least a tacit recognition of the possible gains obtained by agricultural use of waste heat from power plants. Although there are many differences among the configurations, their conclusions are similar and tentative. Not many studies attempted an optimization of the agriculture-waste heat utilization in connection with the energy center configuration they developed. One noted exception was the work done by the Agricultural Engineering and Economics Department of Michigan State University for the Michigan Energy Center Study. The conclusions of their study are contained in their final report "Agricultural Uses of Waste Heat"[31] submitted to the Environmental Research Institute of Michigan, October 31, 1975. The following section is excerpts from this report.

Agricultural Uses of Waste Heat at Michigan Energy Centers

General Site Requirements

For sites in the U.S., the general characteristics of the site can produce a significantly different set of uses; for example, MSU[33] suggests that the site be studied from the standpoint of its resources and its needs. For a coastal site in a warm arid region, one mix of uses (desalting, irrigation, greenhouses, industrial chemicals) would apply, whereas for an inland urban site in the northeast, another mix (urban heating, agricultural applications) would be more attractive.

However, much less climatic diversity exists for sites within Michigan. Also, of the many agricultural uses of waste heat suggested in the literature only a few will be economically feasible in some regions. Hence, only four site criteria for agricultural uses seems to be significant for Michigan sites:

1. The projected land use for the area adjacent to the site should be agricultural.
2. The site should be near urban markets.
3. The region should be agriculturally productive.
4. There should be an adequate supply of cooling and irrigation water.

The land area specified for the Michigan energy center does not include land for agricultural use. This land would be adjacent to the park itself. However, it would probably be better if this adjacent land was already agricultural by use. Land currently used for agriculture has the desirable characteristics of being both nonwooded (which minimizes effects on the environment) and level (which minimizes pumping requirements). The land does not have to be prime agricultural land; marginal land can be used and, in fact, may be more profitable, since it can be obtained at a lower price.

The proximity of the park to urban markets is important for fresh market sale of flowers, fish and vegetables. Transportation requirements are not critical; only highway transportation is required for crop transport to markets.

The region near the center should be productive for two reasons: (1) the improvement in agricultural productivity from waste heat should not be limited by climatic factors, and (2) nearness to productive land is desirable to minimize the cost in transporting high-moisture corn to the grain-drying complex at the center.

A source of water is needed both for (a) makeup water lost to evaporation and seepage into the soil and (b) water to meet conventional cooling system requirements after the waste heat water leaves the biological subsystem and before it returns to the power plant condenser.

The energy center sites studied for Michigan are near Harbor Beach and near Rothbury-Montague. The Harbor Beach site, east of Saginaw, has the following characteristics:

1. It is in a region that is, and will probably remain, rural.
2. It is close to the major grain-producing area of Michigan. The site and surrounding area is currently used for cash crops (corn, soybeans, sugar beets). A site is currently being selected for a grain elevator complex in the region.
3. The soil is generally high-to-medium-fertility loams and sandy loams (Kawkawlin Conover). The topography is flat plains.
4. The area is adjacent to and could obtain cooling water from Lake Huron. There are some drainage problems reflected by the presence of a network of farm drains.

The Rothbury-Montague site near Muskegon has the following characteristics:

1. The southern border of the site will encounter land pressure due to urban growth. Some residential development exists about a mile south of the site.
2. The agricultural productivity is limited by the infrequent occurrence of fertile soils. The site and surrounding area is currently a mixture of forest and cropland. The soil is generally fine sandy loam of moderate fertility (Bohemian Selkirk). The land is gently rolling.
3. The site is adequately served by interstate highways to markets both in Chicago and in major population centers in southern lower Michigan.
4. The site is adjacent to and could obtain cooling water from Lake Michigan.

As noted, power plants for electric generation can achieve efficiencies of only 30-40%. Thus, a Michigan energy center generating

23,125 MWe of power could be expected to discharge around 140 x 10^9 Btu/hr of waste heat to the surrounding environment. An integrated agricultural system has been shown to be a least-cost system among current alternatives (wet-draft cooling tower, dry-draft cooling tower, spray canals and cooling reservoir) for dissipating the excess heat. Utilization of waste heat in this manner achieves significant social benefit to the local and state-wide economy.

This discharge of heat is subject to legal regulations in order to ensure environmental quality. Usually the heat must be dissipated in cooling towers or reservoirs. The conventional methods of dissipating thermal discharge are a waste of both capital and potential energy resources. Agricultural systems are a promising alternative but only if they are integrated into the other systems of the center. This is because an engineering analysis of single-use systems indicates that the beneficial uses of discharged waste heat from steam electric power plants will not have any effect on alleviating potential thermal pollution problems of steam plants, or significantly aid in resolving power plant siting problems in the future.

The Michigan State University study developed a system comprised of 100 ac of soil warming, 500 ac of fish-rearing facilities, 300 ac of greenhouse, 65 ac of grain drying and 85 ac of waste treatment. Not all of the promising uses are suitable at a given site. The biocomplex selected for the Harbor Beach site is a combination of fish culture, grain drying, waste treatment and greenhouse warming. The biocomplex for the Montague site is a combination of fish culture, greenhouses, soil warming, irrigation and sport fishing. The combination of uses selected was determined by balancing higher costs for nonideal uses against the need to have a fairly constant nonseasonal demand for the waste heat. A constant demand will minimize the capacity of the needed supplemental conventional cooling and thus reduce the nonproductive capital costs. The waste treatment facility at Montague was excluded because there already exists a large waste treatment plant at Muskegon. Grain drying was excluded at Harbor Beach because it is not in a corn-growing region.

The biocomplex description includes specifying the beneficial uses, their sizes, production capabilities, water requirements, capital requirements and limitations. These factors are presented in Tables XVIII through XXI.

These tables note that a single siting would use approximately 12.8% and the dual siting would use 11.3% of the total waste heat available. Total heat dissipated is about 20 x 10^9 Btu/hr and the consumptive water use is about 27 mgd. Possible limitations on size are market constraints, managerial ability, diseconomies-of-scale, and availability of waste heat. It is clear from the above analysis that the availability of

Table XVIII. Percent Waste Utilization and Size for Alternative 23,125-MWe Energy Centers

	Site 1 23,125 MWe		Site 1 15,000 MWe		Site 2 8,000 MWe	
	%	Area (ac)	%	Area (ac)	%	Area (ac)
Fishery	8.7	500	5	300	3.3	200
Greenhouse	2.8	300	1.8	200	0.9	100
Grain Drying	0.7	65 (10 Mbu)	0.7	65 (10 Mbu)	N.A.	N.A.
Irrigation	N.A.	N.A.	N.A.	N.A.	0.1	3,200
Recreational Fishing	N.A.	N.A.	N.A.	N.A.	N.A.	N.A.
Soil Warming	N.A.	N.A.	N.A.	N.A.	0.1	200
Waste, Secondary	0.0	10 (10 mgd)	0.0	10 (6.2 mgd)	N.A.	N.A.
Waste, Tertiary	0.6	85	0.4	52	N.A.	N.A.
Subtotal	12.8	960	7.9	627	3.4	3,700
Total	12.8	960	11.3	4,327		

Note: N.A. = not available.

Table XIX. Biocomplex Energy Center Components

Use	Size (ac)	Heat Dissipated (10^6 Btu/hr)	Water Used (mgd)	Capital (million $)	Revenue (million $)	Limit
Fishery	500	13,000	1.6	4.6	1.3	Market
Greenhouse	300	4,200	0.1	33.0	9.0	Market
Grain Drying	65	1,100	Nil	20.0	1.4	Location
	(10 Mbu)					
Irrigation	3,200	210	25	1.6	0.03	Soil type; Crop
Recreational Fishing			Once-through		No estimate	EPA
Soil Warming	200	110	0.6	1.0	0.04	Polyvinyl Chloride
Waste, Secondary	(10 mgd)	6.1	NA	2.2	0.04	Population
Waste, Tertiary	85	1,000	0.3	0.3	0.13	Secondary

Table XX. Initial Capital Costs, Annual Operating Costs and Pumping Costs per Subsystem[a] (Millions of Dollars)

Subsystem	Initial Capital	Annual Costs		Pumping Costs	
		Annual	Discounted	Annual	Discounted
25 20-ac Fish Ponds	4.6	2.3	15.0	1.3	8.5
200-ac Soil Warming (Vegetables)	1.0	0.4	2.6	0.02	0.13
300-ac Greenhouse	33.0	51.0	333.0	7.5	49.0
100-Mbu Grain Dryer	20.0	3.0	19.6	4.0	26.1
85-ac Algae Pond	0.34	0.12	0.78	0.02	0.11

[a]The annual operating costs and pumping costs are shown in annual figures and for the corresponding net present value of those annual costs at 15% interest for 28 years. These figures are exclusive of the overhead costs associated with the general piping and distribution system. Annual operating cost figures exclude pumping costs.

Table XXI. Resource Requirements and Annual Income Statement for the Specified Subsystem Design (Millions of Dollars)

	Soil Warming (Vegetables)	Green-house	Fish Ponds	Grain Dryer	Algae Ponds	Total
Initial Capital Requirements	1.0	33.0	4.6	20.0	0.34	58.94
Annual Gross Revenues	0.44	67.5	5.0	5.6	0.15	78.69
Annual Operating Costs	0.40	51.0	2.3	3.0	0.12	56.82
Annual Pumping Costs	0.02	7.5	1.3	2.5	0.02	11.34
Annual Net Returns	0.03	9.0	1.4	0.1	0.01	10.54

waste heat is not a dominant factor since only 12% can be diverted to agricultural uses. Rather it is the market constraints and, perhaps more importantly, adjacent land constraints. In the latter case, it is necessary to have the applications of waste heat located very near the generating facilities since it is expensive to pump the warm water long distances. The pumping costs soon outweigh the advantages when even moderate distances are considered. In addition, there is strong competition for land adjacent to an energy center for additional industry and population centers; this will require trade-offs which will affect the feasibility of the entire concept. It will be shown later that this limitation on the capacity of agri-aquacultural applications may well dictate smaller energy centers at some locations than currently proposed.

Summarizing, one finds that agricultural uses of waste heat from power plants and energy centers is an economically attractive concept, but one which has limitations. It is also imperative that such applications be matched and optimized along with the rest of the system.

TRANSPORTATION DEPOTS AND PORT FACILITIES

Although most energy center feasibility studies did not include detailed plans for transportation facilities, it is of obvious importance. Two programs, Puerto Rico and Michigan, included consideration of deep-water ports; the other studies used truck and rail solutions to the transportation problem because of inland locations. Apparently one reason for the lack of detailed analysis of transportation facilities is the problem of tying these facilities into the energy center in a symbiotic manner. It is true that all transportation modes are needed for the various elements of the center, but it is not obvious how a transportation facility would be enhanced by being located near an energy center. In a sense, then, the advantages are one-sided; the center would not reciprocate to the transportation sector advantages which could not be found elsewhere.

Harbor Facilities at the Puerto Rico Energy Center

It was suggested that harbor facilities for the Puerto Rico energy center be located in the southwest corner of the site to take advantage of the naturally occurring water depths. Thus there would be good deep-water access to the sea and, kept within the barrier reef islands, protection against swells and heavy seas.

Berthing facilities were planned to accommodate three classes of vessels as follows:

1. Barges of 10,750 short ton capacity with an overall length of 350 ft and a required water depth of 22 ft.
2. Chemical tankers with 40,000 short ton capacity with a length of 680 ft and a required water depth of 40 ft.
3. Oil tankers of 150,000 dead-weight ton capacity with an overall length of 990 ft and a required water depth of 65 ft.[4]

Michigan Energy Center Port Facilities

The Michigan study team determined that the need for a harbor and port facility at the two sites will depend on which modes of transportation are found to be technically and economically feasible. The two major raw materials to be brought into the Michigan center include coal for the power plants and coal gasification plant and oil for the oil refinery.

These two materials alone would necessitate handling up to 30 million tons per year at the port. In order to visualize this amount of cargo it can be compared to the port of Detroit which handled 30 million tons in 1971 and the Duluth-Superior Harbor which handled 37 million tons in total cargo. Only a portion of the energy center material traffic would be moved by water transportation; the exact distribution of tonnage among the various transportation modes was not determined by the Michigan study team as it requires a very detailed technical and economic analysis.

An artificial harbor would have to be constructed at both the Harbor Beach and Montague sites. It was recommended that the harbor be designed to accommodate 1,000-ft ships (such as the U.S. Steel CORT, or the Presque Isle 1,000-ft tugbarge). Such ships have a displacement of more than 100,000 tons and a 30-31-ft draft. At the present time, ships operating in the Great Lakes have a 27-ft draft in order to navigate the limited connecting channels such as the Detroit and St. Clair Rivers. The energy center harbors should be maintained at 28-29-ft depths. A harbor designed for these large ships would also be able to handle the 700-ft ships entering from the St. Lawrence Seaway.

The Michigan study team considered the desirability of providing protection for ships from wave action in the harbor by dikes or break-waters. The size, cross-sections and construction materials for these protective features, and the water depth in the harbor, are determined by considering the wave climate (that is, the average height, direction, frequency and water-level fluctuation over 20 years). For breakwater or dike construction, rubble mound has certain favorable characteristics compared to sand- and gravel-filled steel cylinders. Rubble mounds are easier to maintain, provide a good fish habitat and are better at damping waves and reducing wave reflections, thereby resulting in less scour and deposition.

The harbor has an important effect on the littoral drift along the shore-line. If there is a pronounced net drift, the harbor structures will cause sand to pile up on one side and cause beaches to be gouged on the other side. The Harbor Beach site of the Michigan study is not subject to large amounts of littoral drift because of the rocky shoreline. The littoral drift at this site is estimated to be from 15,000 to 20,000 yd^3/yr. Although the site near Montague on Lake Michigan is much less rocky, it may be close to a littoral nodal zone (where incoming drift rates are equal in both directions and cancel each other). The net southerly drift has been estimated to be 21,000 yd^3/yr. Both sites, then, have a relatively small drift potential.

The Michigan study concludes that dredging will generally be required to keep the harbors navigable. This problem is usually solvable, although at some expense. Previous experience in the Harbor Beach area indicates that some shoaling problems may be expected. On the other hand, the dredged materials can probably be used to replace littoral drift without environmental damage since the harbor is not located near a river outfall and will not be polluted.

The 40- to 50-ft bluffs at the Montague site may cause some problems. These bluffs will interfere with the construction of docking and other port facilities and with the installation and use of conveyors to transport materials to the energy center. Also, moving the unstable clay and sand materials to construct the harbor could cause problems. Constructing an artificial shoreline would be required to avoid problems with the bluff, but could be accomplished if the site was attractive otherwise. In summary, constructing a port facility at either Michigan site will have both negative and positive environmental impacts. Although this was not a detailed study, it would appear that the benefits would predominate.

Being located on a Great Lake, it seems that both Michigan study sites would be well disposed to harbor and port facilities. However, it is instructive to compare these potential facilities with the alternative transportation mode: the railway. For this comparison, it will be sufficient to compare only the requirements for handling coal movement.

At the present time Michigan has an installed capacity of coal-burning power-generating facilities of around 12,000 MWe, which produce about 60,000 million kWh of electricity. This requires 20 million tons of coal per year. If, for the moment, we assume that all new capacity at the Michigan energy center consists of coal-fired units, then approximately 40 million tons of coal are required each year. If this coal were all rail-transported, there would be over 4,000 unit trains making trips to Michigan per year. This is over 10 unit trains coming into Michigan every day of the year. Detroit Edison is currently planning a coal dock facility to handle 20 million tons per year—or about half the energy center needs in the year 2000—at a cost of $50 million. Port facilities at the Michigan energy center would be expected to cost around $100 million in today's dollar. Yet the port alternative may well be preferred to rail transportation for a number of reasons, including the shortage of hopper cars, the greater energy requirements of rail travel, and the problem of providing and maintaining rail right-of-way.

VI

THE ENERGY CENTER AND THE NEW TOWN

There seems to be something in each of us that stirs at the possibility of creating a new community from scratch. Perhaps it is the desire to create, or to control or to determine. Perhaps it is the desire to begin all over again and get it right this time. Or perhaps it is simply the desire to make a profit. Whatever the reason, the idea of creating a new town has an increased scope when combined with the concept of an energy center. Beginning a new town from scratch at an energy center provides many more alternatives for energy conservation and effective utilization than can be obtained with most current cities. In the opinion of some, the new town may also provide many more alternatives for worsening already bad urban situations. It will not be possible to discuss all the fascinating possibilities in this book; however, some of the more important possibilities and problems will be mentioned.

OBJECTIVES OF THE PLANNED COMMUNITY

Broadly stated, the goal of the deliberately planned city is to avoid the evils of modern urban life and to achieve those values that the modern city neglects. Implicit in such a goal is the hypothesis that both the evils and values can be determined, agreed upon and altered. Since we are not interested in these more philosophical points, but rather in how such a new town will fare in conjunction with an energy center, we assume that such goals can be achieved and offer as proof, the continued existence of planned communities. Such communities as Levittown, Columbia, Maryland, Winona, Minnesota, whatever their failings, will serve as proof of the possibility. The goals and objectives of some planned communities will be examined and to the list will be added energy conservation and effective energy utilization. It will be important to see how the addition of newer energy goals may come in conflict with some of the traditional new town

115

goals such as the "return to nature" or "economic self-sufficiency" which motivated some backwoods utopias. The energy center concept may also be somewhat in conflict with the Winona experiment which emphasized energy conservation but without large-scale central generation.

In an attempt to reduce the broadly stated goals of the new town to a number of specific objectives, it has not been possible to remove all ambiguity.[34] However, the goals may include: (1) love and respect of the land; (2) continuation of tradition; (3) economic self-determination; (4) political autonomy; and (5) shared values. Through the achievement of these goals, it is presumed urban conflict, delinquency, crime and pollution will be reduced. It may also be possible, by adding a few more goals, to achieve more effective fuel usage.

Columbia, Maryland—A New Town

The so-called energy crisis was still ten years off when James W. Rouse, a Baltimore banker and developer, purchased about 15,000 acres of land between Baltimore and Washington in 1962. After an 18-month planning period, construction began in 1966; the first residents moved into the new town in 1967. The growth of Columbia has been rapid: from 1,300 in 1968 to 25,000 less than five years later. It is expected to have a population exceeding 100,000 by 1981. At that time, it is planned that Columbia will consist of seven villages of 15,000 residents each. Each village will contain commercial facilities, a secondary school and recreational areas. Each village will consist of four or five neighborhoods with their own elementary schools within walking distance. The neighborhoods and villages are linked together by a town center with a shopping center and office building complex. Recreational activities are centered around Lake Kittamuqundi. Industrial development has proceeded rapidly in the form of several research and development firms sited at the Oakland Ridge Industrial Park and the Guilford Industrial Park. General Electric began operations in 1971 on a 1,100-ac site employing approximately 3,000 workers. The industrial park employment associated with the development of Columbia is expected to total 12,000 by 1980.

Columbia is distinguished not only by its rapid growth but, more importantly, by its commitment to a communitarian ideal, that is, by sharing common values and goals. In this respect, then, such a new town has many characteristics of the new residential areas envisioned for the energy centers, although some of the goals and values have been shaped and rearranged by the events arising from the energy crunch. Columbia is proof that careful planning and development can create a new town from nothing to a 100,000 population in twenty years. This is the same growth rate projected for the Michigan energy center; it seems an achievable goal. Columbia, then, can serve as a model (requiring some revision) for the cities projected for inclusion in the typical energy center concept.

In Columbia, efforts have been made to attract and achieve social integration with respect to race, religion, age and wealth, characterized by its reduction of barriers between the community and its institutions. A variety of religious facilities, several colleges, moderate-income housing and high-rise apartments have been established which leads to the realization of an "open-door" policy.

The history of new towns abounds with rationales for their plans or their existence. A partial list would include[34]:

1. A vehicle for a national growth policy.
2. A vehicle for innovation.
3. A means for a more pleasing environment.
4. A vehicle for effective social reform.
5. A source of profit for developers.

Investigations into the history of new towns may show somewhat different lists; however, few if any list has the conservation of energy as an important goal. Although a detailed analysis of energy conservation at a new town cannot be performed in this book, it will be argued that energy considerations should be primary in future new cities, especially those associated with energy centers. It is only possible at this time to pose some questions to answer when considering a new town at an energy center.

Is energy conservation a valid consideration for a new town? Is a new town compatible with the objectives of an energy center? Is the energy center concept compatible with a new town? Can we determine purposes for the town and energy center that are meaningful? How can we determine the proper size for a new town associated with a given size energy center? Will the political body which has control over the energy center also have control over the city? Why or why not? Are new and different relationships among the people, industries, the local government and the utilities necessary? How will shared services (heat and power) be administered and paid for? How will assessments and taxation be levied between industries and the community? How will the supplied services to the new city be financed? Who is to decide on the level of services provided? How are rates for services to be determined? Should all social services (schools, libraries, health care facilities, law enforcement, etc.) for the new town be planned and coordinated or only the utilities (heat, power and water)? Will a population attracted to a planned community be the same population needed for the proper operation of an energy center? Or are these "different kinds of people"? Will such a planned community be criticized as too "socialistic"? Can such charges be refuted or are they true? And will the overall expected advantages outweigh the necessary increase in effort to achieve desirable objectives?

The experience of Columbia, Maryland, shows that many of these questions cannot be taken lightly. Although it may reasonably be considered that Columbia is a success, many of the original communitarian social values were rejected or at least could not be agreed upon among residents, experts and developers. There were also many instances of community conflicts in the new town. It can well be expected that there will be disagreement and conflict concerning the objectives of an energy center town also. Why did some of the planning fail at Columbia? Can an energy center benefit from the experience? Can it be that organized community planning to achieve social and economic goals is neither necessary or possible? It is also quite likely that the eventual residents, those most affected, will not have a voice in much of the planning. For example, at Columbia the developer had to secure new town zoning before any residents arrived; consequently, no new town residents were involved in the initial structuring of the ordinance.[34] This meant that much of the structure of the new society was quite dependent upon specialized and technical personnel servicing the developer. Such a circumstance will also be expected at an energy center and will result in some conflicts and the alteration of some goals and objectives. This may not be a fatal flaw for new towns but it does indicate the complexity of the problems that must be faced along with many others at an energy center. Many of the studies of new towns emphasize the desirability of locating residences away from heavy industry in order to enhance human values which many see as incompatible with such endeavors. On the other hand, the concept of a new town at an energy center is directly in opposition to much of this ideal. It takes little imagination to realize how much more difficult it will be to organize a new community at an energy center.

Before developing a pessimistic tone, it will be useful to describe some of the expected advantages for locating a new town near an energy center:

1. Utility services, although not commanding a lower rate, will nevertheless result in lower costs when delivered to nearby communities.
2. Economical district heating of residences and commercial buildings is feasible because of the close distances and because only new construction can take advantage of this source.
3. The transporting of the heat for district heating may take place under streets and sidewalks, reducing the need for snow removal in some localities.

4. Solid waste and sewage disposal for the new town can be incorporated into the energy center facilities to not only serve as an additional source of energy but to reduce the magnitude of the disposal problem. The usual pollution problems may thereby be alleviated.
5. Public transportation should also be considered anew because of the possible feasibility of electric buses or an equivalent.
6. The center provides a stable job market and the community provides a stable work force.
7. Hot water for home and commercial use can be provided by the district heating system as well as energy for air-conditioning.

Although energy conservation and other energy considerations were not a primary impetus in planning Columbia, many of the experiences obtained would be helpful in designing a new town at an energy center. Rather than passively awaiting the emergence of typical urban problems that often occur with the growth of cities, Columbia has a degree of order, stability and opportunity unmatched in any growing area of the country. These potential advantages are available in almost every element of the environment: shopping, entertainment, libraries, health, communication, transportation, culture, safety, institutions, industrial and business development and, of course, energy.

During the initial planning phase of Columbia a number of different energy systems for the new city were investigated and considered by Albert B. Gipe and Associates.[35] Since this was to be a completely new, modern city, analysis and investigation was performed by Gipe to determine the most up-to-date and advanced energy systems available at present or within the near future. These included nuclear power generation, on-site generation, purchase in bulk and resale of electrical and gas services, central distribution systems, and central heating and cooling plants utilizing various fuels.

Initial studies indicated that it would be desirable to purchase electrical power in bulk from the local utility company which served the general area, to install, operate and maintain an electrical distribution system, and to resell electric power to the various consumers. It was determined that on-site power generation would not be economically feasible because of the high initial capital investment and because the load would build up over a period of years. Of course, at an energy center there is an original commitment to construct power generation to supply a large region.

Initially it was planned to purchase, distribute and resell power to the users at Columbia. Eventually negotiations were directed toward a transmission and distribution system to be furnished and installed by the Baltimore Gas and Electric Company. The general arrangement of the distribution system supply and overall appearance of the substations and distribution equipment were to be approved by the developer. Albert B.

Gipe and Associates were assigned the task of coordinating and reviewing the electrical energy and system requirements with the local utility.[35] Similar considerations were given to the study of gas and fuel oil distribution.

Feasibility studies were made for various central (district) heating and cooling systems utilizing oil, gas or electricity as the prime fuels. In the more densely populated areas and in areas of higher load density, the studies indicated that central heating and/or cooling would be feasible. Consideration was given to the use of central plants in various sections of the development.

Of interest to energy center planning is the prediction of the electrical loads for the city, and the staging of these loads not only so that planning of the electrical distribution system could proceed, but also to arrive at equitable cost and charges for the electrical distribution systems. It had been decided at the beginning that the entire electrical distribution system would be underground. All transmission and primary distribution cables would be underground, as well as all secondary, main and service cables.

Since Columbia is expected to have a population of around 100,000 persons by 1980, it may be possible to project some of the experience gained in the energy area to an energy center new town of from 100,000 to 200,000 population by the year 2000. A projection for power requirements for Columbia for 1980 is similar to the following (Table XXII):

Table XXII. Residential, Commercial and Industrial Power Loads for Columbia, 1980[35]

Residential: 30,000 dwelling units	
Lighting and miscellaneous at 3 kW	90,000 kW
Air conditioning at 3 kW x 60%	54,000 kW
Electric heating at 20 kW x 20%	120,000 kW
	264,000 kW
Town Center Commercial	
Lighting and miscellaneous	35,000 kW
Air conditioning	25,000 kW
Electric heating at 20%	10,000 kW
Miscellaneous	5,000 kW
	75,000 kW
Village Commercial	
Lighting and miscellaneous	5,000 kW
Air conditioning	4,000 kW
Electric heating at 20%	2,000 kW
Miscellaneous	1,000 kW
	12,000 kW
10 villages =	120,000 kW
Industrial	100,000 kW
Miscellaneous	5,000 kW
Total projected connected loads	564,000 kW

Based on the Columbia experience, it may be projected that a city of 100,000 at an energy center would require from 200 to 500 MWe of power from the generating facilities, not including the need for industrial users. The lower figure would be closer if district heating or residences and commercial building predominated, while the upper figure would be expected where significant electric heating prevailed. This projection seems to compare closely with the independently obtained estimate for the Michigan energy center city requirements of 238 MWe annual average electrical power and 240 MWt annual average district heat production.

Urban Development at the Michigan Energy Center

The Michigan study team determined that to meet the needs of the temporary and permanent employees at the energy center, as well as to provide for associated economic activities (other than the heavy industry located at the industrial center), it would be necessary to provide large-scale residential, commercial and institutional facilities within reasonable distance from the center site. The Michigan concept provides for a center located in an area with an initially low population density. Therefore, new facilities will be required which would be most economically provided by the construction of a city primarily oriented toward meeting the needs of center activities.

The center industries chosen for the Michigan study were generally capital-intensive so that the total number of center workers would be less than suggested by the total capital investment. On the other hand, it is likely that a broad diversity of social and economic functions could be provided by the city, including many that are only indirectly related to the park objectives. Thus, the center city could serve a number of purposes:

- It would provide residences for center employees, employees of the service facilities and industries, and their families.
- It would provide social services (schools, fire and police protection, hospitals, libraries) and shopping areas for the city population and for the surrounding areas.
- It would provide the marketing, financial and business activities needed for the center industries.
- It would provide commercial and industrial facilities not necessarily associated with the energy center.
- It could provide many of the communitarian goals which were important for Columbia and other planned communities, including the desire for a more pleasing environment, social integration and a vehicle for achieving shared values.

The Michigan team assumed that the population of the city would eventually grow to be 200,000 sometime after the year 2000. This estimate was used for determining the land, water, energy and other services to be provided. The city planning should also require design features that emphasize energy conservation. As a model for such a city, Michigan used the general features of a reference city discussed by Miller *et al.*,[39] that makes extensive use of heat energy from a steam-electric power plant for space heating, air conditioning, water heating and sewage treatment. A further discussion of this reference city will be provided in the next section.

The Michigan city is to have a high population density, with its residential areas primarily composed of three-story apartment buildings. Adequate space must be provided for social service institutions such as schools, hospitals, churches, government offices, recreational facilities and shopping areas. The high population density should result in efficient distribution and use of energy if adequate planning and appropriate design is undertaken.

District heating and air conditioning are already in use in many large cities. These systems have been in operation for many years, providing a large experience base for determining relative costs. The economic justification for a particular installation must be determined from data for the particular areas served, the climate, distances from the power plant, capital costs and alternate fuel costs, but it appears that district heating and air conditioning will often be fully competitive with other sources of energy. Thus the future use of steam or hot water for energy distribution will enable one to substitute coal and nuclear fuels for oil and gas. From the standpoint of energy conservation, the district heating system would have the same general advantages as other uses of process steam for the increased overall efficiency of the combined generation of electricity and steam. To a lesser extent this is true for air conditioning also.

In the Michigan system, the energy will be supplied to the city through a distribution system carrying 300°F water. The air conditioning is to be produced using 2-psig lithium bromide-absorption refrigeration equipment. The peak flow rate for this system occurs in the summer when 6,500 lb/sec of 300°F water is used to supply the summer peak air conditioning load.

Table XXIII summarizes the data on the city and its energy use patterns and costs.

Table XXIII.   Center City Statistics[1][a]

Total population	200,000
Population served by district heating system	135,000
Area served by district heat system (mi^2)	10
Population density in 3-story apartment area (people/mi^2)	20,000
Capital cost of district heat distribution system	$46 million
Hot water cost	$0.90/MBtu
Distribution cost for hot water	$0.90/MBtu
Annual average net electrical power	238 MWe
Annual average district heat production	240 MWt

[a]Costs are based on assumed completion in 1980.

PROVIDING THERMAL ENERGY TO URBAN AREAS FROM ENERGY CENTERS

Miller *et al.*[36] have shown that with coordinated planning of energy centers and new cities, it would be feasible to provide thermal energy from steam-electric power plants to urban areas. Their analysis was made for a reference city of 300,000–400,000 people by the year 1980. Thermal energy from the power generation facilities would be used for space heating, hot water and air conditioning for both the commercial buildings and most of the city's inhabitants. Apartment areas in the reference city were considered to have an average population density of around 21,500 people/mi^2. Heat would also be available for manufacturing processes and for sewage plant use. This study estimated that this use of heat from the generating facilities would reduce cooling tower capacity requirements by around 37% for a single unit. The heat rejection to the environment would be reduced to 21% of that from a single-purpose plant during the period of maximum heat consumption in the summer. Of course these percentages of waste heat usage would not be obtained at an energy center since the ratio of generating capacity to city population is projected to be much more at the center.

The cost of distributed hot water for the reference city was calculated to be around $1.42/MBtu, which compares favorably for most U.S. cities. The cost for both space heating and domestic hot water would be around $1.98/MBtu. The study further notes that if the plant cooling water were used to heat and air condition greenhouses, the cooling towers would be eliminated and, with no charge for greenhouse heat, the cost of heat for the city would be slightly reduced. It was also determined that heat from the generating plants might be used for urban vehicle propulsion and snow melting.

The Oak Ridge National Laboratory team which conducted this study was convinced that it showed the feasibility of serving new cities with heat from a central power station, thus reducing air and water pollution. The study showed that for many heat-consuming processes, the use of rejected or extracted heat would conserve all or part of the fossil fuel that would otherwise have to be burned in order to provide heat. Such uses of heat would eliminate the atmospheric pollution from gases and particulates caused by the additional combustion process. It would also reduce some of the accompanying heat addition to the biosphere. The heat rejections at the condenser from single-purpose generating installations employing fossil fuel and nuclear fuel range from 53 to 66% of the energy input in the turbine.

The reference city used for the Oak Ridge study is shown conceptually in Figure 16. The residential and commercial areas of the city are all situated at least five miles from the energy facilities. The downtown area and an apartment house area are between 6-12 mi from the power facilities and receive 300°F water for building services. This section of the city has a total area of 16 mi^2.

The Oak Ridge study considered many of the same uses of waste heat and process steam as did the other studies, including greenhouses, sewage treatment, district heating and cooling. However, it included another possible use not considered in other studies: urban vehicle propulsion with hot water. A study at Oak Ridge has shown that the heat distributed through a district heating system can be used to reduce the air pollution by providing energy for buses or trucks in the form of stored steam or superheated water. Such an arrangement has already been in use, most notably for applications which must avoid sparks such as powder plants, chemical plants and tobacco warehouses. It could also be used for urban public transportation.

At one time the automotive steam power plant was a serious competitor of the internal combustion engine. Once again, the steam engine should be reexamined in view of the rapidly growing problem of air pollution. To avoid venting steam to the atmosphere a closed-cycle engine could be used.

The energy would be stored in tanks in the form of superheated water rather than pressurized steam. This would provide about ten times more energy per cubic feet of tank capacity. The amount of energy that can be stored per pound of storage medium or per cubic feet of volume is the limiting factor in the usefulness of any stored energy system. By way of comparison, the useful energy available at the drive wheels of a vehicle using gasoline is about 3,600 Btu/lb of fuel. Superheated water, at 400°F released from 1,000 to 50 psia, provides only 58 Btu/lb which,

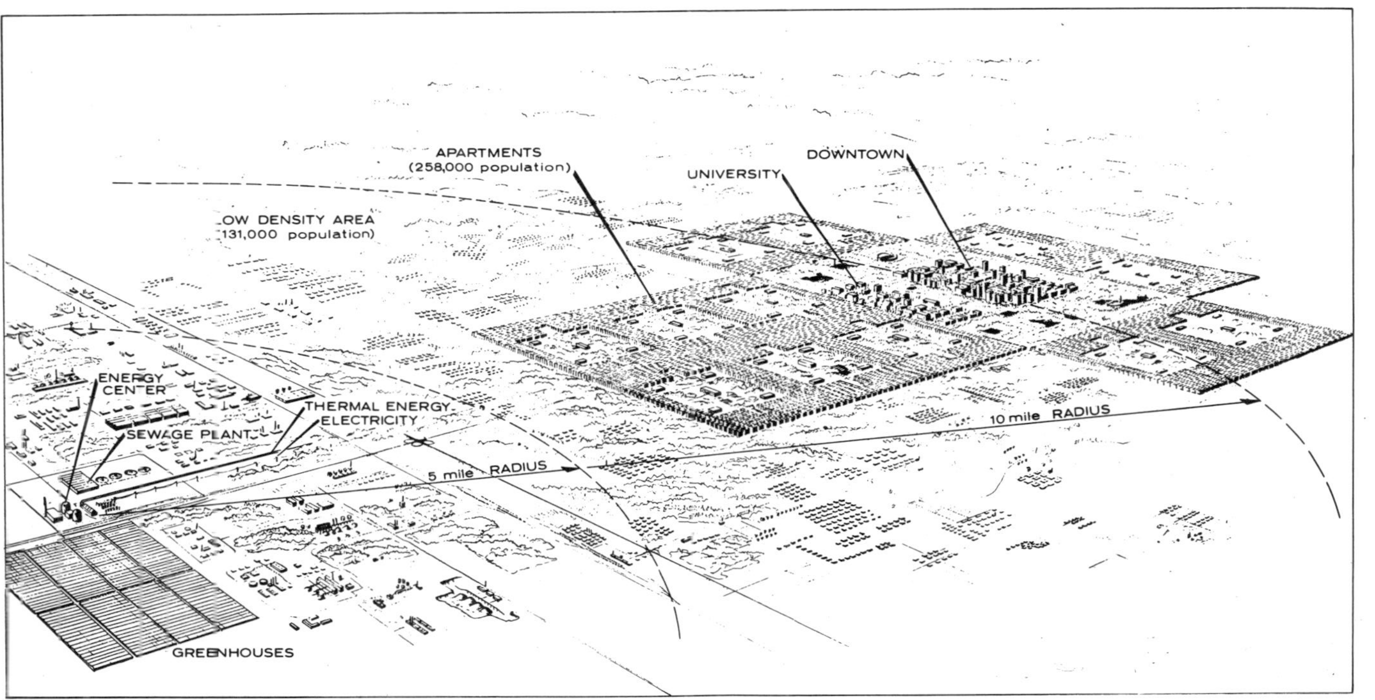

Figure 16. The reference city used for the Oak Ridge study.

in a manner, demonstrates the popularity of using gasoline in internal combustion engines. On the other hand, superheated water compares favorably with lead-acid and nickel-cadmium storage batteries at 41.5 and 35.2 Btu/lb of useful energy, respectively.

Vehicles that use energy storage systems, such as batteries or superheated water, are limited in their range of operation by the space and load capacity available for the energy storage units. Besides being non-polluting and using superheated water readily available from an energy center, the steam-powered vehicle has other advantages. For one, the specific fuel consumption at part load is less for a steam engine than for either gasoline or diesel engines. This is because gasoline and diesel engines have a marked increase in specific fuel consumption if the engine is throttled, as is necessary at reduced loads. In addition, there would be no heat losses in the engine due to combustion effects with steam engines and there would be no losses in the transmission because a direct-drive steam engine would give an exceptionally smooth fast start.

The operating range of a superheated water vehicle can be estimated by assuming that 20% of the gross vehicle weight would be devoted to tankage and that the weight of the tanks would be 25% that of the contained superheated water. Under representative load conditions, it would require 120 pounds of superheated water per mile. A 16,000-lb bus in which 20% of the gross weight was allocated to the energy storage system, would have a range between refills of approximately 20 miles. For urban transit systems, such a range would be adequate. The comparative costs of such a transportation system were not estimated in the Oak Ridge study. They would, of course, depend on the configuration of the adjacent energy center.

The use of hot water circulating beneath sidewalks and roadways for snow and ice removal was considered by the Oak Ridge study. A rather elaborate arrangement of pipes, valves, heat exchangers and control systems was described which delivered 107 Btu/ft^2; of area and had an annual cost of from 39 to 72 ¢/ft^2 of heated area. For roadways this could be an annual cost of $100,000/mi and would not be practical, except perhaps for airport runways where safety considerations would warrant high costs. Perhaps a simpler scheme would be more suitable. For example, it might be possible to direct pipes carrying district heating water or waste hot water from the power plants' condensers under roadways and sidewalks. This may provide only partial snow or ice melting, but a reduction in the pipe insulation would help offset costs due to increased heat losses. Many college campuses use this method of tunneling heat along thoroughfares, giving good snow removal at nearly no additional cost. Severe snow accumulation during winter storms could be removed using conventional methods and dumped at sites nearer the energy center where waste heat could be used for melting. The warm water from the condenser of the energy center offers a free heat source for this purpose.

SUMMARY

Although many serious problems will attend the construction and habitation of a new city at an energy center, most studies, analyses and experiences have shown that such a city is feasible and that there may be desirable reasons for its creation. A common motivation present in studies of new towns is the hope that careful planning will allow the job "to be done right this time." Constructing a new town near an energy center allows all elements of the entire complex to take advantage of several symbiotic relationships, the most important of which may be the potential for energy conservation. Exact final savings cannot be determined in the studies; however, it appears that savings of anywhere from 10 to 50% of the heat content of the fuels may be saved while improving the quality of life for the city inhabitants. Only cursory examinations of new towns at large-scale energy centers have been performed; the initial conclusion is that it would be a feasible and probably advantageous combination.

VII

ENERGY CENTER REQUIREMENTS

This chapter discusses water, climatological, geological, transportation and labor requirements for a large-scale energy center of a capacity of around 24,000 MWe. The data are drawn largely from the Michigan example.

LAND AVAILABILITY

The land requirements for an energy center should be specified both in terms of type of land and required areas. The land should have a relief similar to that required for any power-generating facility and should have good drainage. Nonforested areas would reduce the initial land preparation costs. If nuclear power generation is to be included, the existing population densities within and around the site should be low to minimize the cost and effect of relocation and to meet Nuclear Regulatory Commission guidelines. Sites should have acceptable access and linkages to transportation facilities. Distances to markets for the associated industrial products and sources of raw materials must be considered. Since large-scale generating facilities are to provide power for a region, the distance to these load centers must be economically reasonable. Transmission corridors must be available. Proper attention must be paid to the availability of dedicated properties and public lands, such as recreational lands, schools, churches, cemeteries and historical and scenic areas. If agricultural activities are to be associated with the center, sufficient productive land with low relief must be available. Because of the large water requirements for cooling and industrial processing, the site must be near a large reliable water source not dedicated to other purposes.

Sufficient land must be available for the power plant, associated industries, agricultural activities, center expansion, transportation routes, and for the residential, institutional and commercial activities of the population to construct and operate the center.

WATER REQUIREMENTS

For a 24,000-MWe energy center, the water requirements are so great as to necessitate a coastal or near-coastal site. No groundwater source would be adequate to meet the water needs and only the largest rivers whose waters are not already intended for other needs could be used. In the Michigan study, cooling water makeup supplies must be obtained from the almost unlimited heat sink of the Great Lakes. It was determined in the Michigan study that the distance from the Great Lakes at which the cost of cooling towers is balanced with the cost of structures for once-through cooling is about five miles. The boundary of the center was set at no more than two miles from the shore.

The Michigan energy center as described would require a large amount of water, primarily for heat dissipation and industrial processing. It was assumed that all industries would operate on a closed-cycle water system so as to minimize consumptive water demand and the environmental impact from wastewater disposal. In this case, the total industrial consumption of water would be about 52 mgd, or about 81 ft^3/sec. Most of this water would be used as makeup water, and so this amount would be released to the atmosphere. The water requirements for the energy-generating portion of the center would be about 170 mgd for the condition that 5×10^6 lb/hr of process steam would be utilized by the industrial portion of the center.

There is still some question about which method of cooling should be used for energy centers. In the Michigan case, the center description was based on the use of evaporative cooling towers, primarily because of possible dangers from ejecting warm effluent directly into the Great Lakes. Alternate methods of cooling were also given brief consideration, in spite of possible other disadvantages. The use of cooling ponds would greatly increase the required center land area, while once-through cooling would probably be restricted by environmental regulations. Because of increased evaporative losses, the use of cooling ponds would increase the consumptive use of water for cooling purposes from 170 to 212 mgd. The use of once-through cooling would decrease consumptive use to 119 mgd, but the gross flow of water would amount to 19,500 mgd. A mix of the three methods might be a workable compromise which would result in consumptive water use of 167 mgd and a gross flow of about 6,600 mgd.

CLIMATOLOGICAL REQUIREMENTS

The climatic requirements for an energy center would not be substantially different from any power plant except as the climate may affect the potential for air pollution. Thus areas should be avoided which have a high frequency of atmospheric inversions or stable conditions that would tend to suppress atmospheric dispersion. Energy center sites must be selected which can allow the federal, state and local ambient air quality standards to be met. The main difference between the energy center and the usual single or small group of power plants is that the air pollution potential is greatly increased over a smaller area. This can be viewed either as an escalated threat or as an opportunity to apply more ambitious methods of pollution control.

Another climate requirement for energy centers is a low frequency of severe weather conditions, such as strong winds, tornadoes, hail and extremes of precipitation. Such conditions could cause more serious difficulties at a center than at a single power plant since several generating units of transmission lines may be affected by a single storm causing outages with a higher percentage of the total load. Areas with a naturally high incidence of fog, or a high probability that the center would create fogging, should also be avoided. Severe weather, fogging or icing would interfere with center construction as well as operation. Any center planning associated agricultural activities must consider the effect of climate on crop growth.

SEISMIC AND GEOLOGIC REQUIREMENTS

Although damaging earthquakes would be very serious for conventional coal power plants, seismic and geologic siting requirements for an energy center will be determined by those criteria that are applied to nuclear power plants. The structures, systems and components important to safety must be designed to withstand the effects of natural phenomena such as earthquakes, tornadoes, hurricanes, floods and seiches. There may also be the need for spacing criteria for the units to prevent interaction.

The Nuclear Regulatory Commission requires an investigation of sufficient detail to provide reasonable assurance of site adequacy. These investigations include determination of lithologic, stratigraphic, hydrologic and geologic conditions of the site and the surrounding region. Also included in these investigations would be the identification of structures, determination of static and dynamic properties of the materials underlying the site and an investigation of all reported earthquakes which have or could have affected the site.

Michigan lies in one of the most seismically stable regions of the U.S. and it is therefore believed that an energy center located in Michigan will be able to meet the two earthquake design criteria:

1. Plant facilities capable of a safe shutdown in the event of the maximum probable event that could occur at the site.
2. Plant facilities designed to incur no damage or loss of function as a result of an earthquake that is likely to occur during the life of the center.

TRANSPORTATION

Extensive water and/or overland transportation facilities will be required at a center for the transportation of raw materials to the center site and finished products away from the site. For the larger energy centers proposed, the transportation requirements for raw materials, especially coal, can be imposing. A discussion of the problems entailed in a total coal-fueled energy economy is outlined in the following section.

When large-scale transportation of major raw materials, such as coal and oil, are required for an energy center, the use of economical water transportation must be considered. Where possible, port facilities must be provided and efforts should be made to expand existing facilities, such as at the Puerto Rico facility. Otherwise, a new port would have to be constructed, as at the Michigan energy center. In addition, a good highway network reaching the center would be required for many purposes, including the transportation of agricultural and fish culture products to market. At some centers, pipelines would be a desirable mode of transportation for liquid fuel products manufactured at the center. Railroad transportation of coal by unit trains would also be an important process for most energy centers.

LABOR AND CAPITAL REQUIREMENTS

Both labor and capital requirements are highly dependent on the size of the particular center proposed. Each study has made its own estimates of these requirements. The labor requirements include the need for construction teams, operating personnel, and the work force necessary for the associated commerical and service endeavors. For those centers which envision the building of an integral new town, the labor needs would be almost completely provided. For those centers which intend to draw labor forces from surrounding populated areas, special problems of dislocation and competition will occur. Because of the scope of most energy center plans, capital requirements appear staggering. This well may be one of the most serious problems facing the planning and construction of energy centers. Capital needs are so great that few private enterprises could cope without joint effort. This, in turn, would open up charges of monopoly and unfair trade practice. Statewide capital formation may also be insufficient. Perhaps only federal involvement could provide the large amounts of capital required, but this alternative would most certainly draw power for decision-making away from the affected populations.

SUMMARY

In the following chapter, we will compare some of the proposed energy centers in their ability to meet the many special needs of large-scale energy generation. These requirements have been summarized for the Michigan energy center in Table XXIV.

Table XXIV. Energy Center Requirements

	Power Plant	Industrial Park	Biocomplex	Center City	Total
Land (ac)	9,510	3,000	960	6,400	19,870
Capital (millions $, 1975)	10,490	1,000	59	–	11,549
Operating Labor	1,600	6,400	300	–	8,300
Consumptive Water	170	52	27	–	249
Heat Dissipation (10^6 Btu/hr, peak value)	117,800	14,000	19,200	2,850	153,850

SOME SPECIAL REQUIREMENTS FOR CONVERSION TO A TOTAL COAL ENERGY ECONOMY

At a time when oil and gas reserves in the U.S. appear greatly limited, renewed attention is being given to the use of coal as a means of providing for our energy needs. Various estimates have placed the known reserves of coal in this country as sufficient to provide energy at current rates of consumption for 500 to 1,000 years or more. What would be the implications of using coal as the total energy source for an energy center? Detailed studies have yet to be made, but much of the impact can be estimated. Some of the implications can be determined by considering a switch in Michigan to a total economy. A brief outline of a study to determine feasibility can also be provided.

Michigan is a state poor in fuel resources: it currently imports 96% of its energy needs. At the present time, less than one-third of this energy is provided by coal. This has not always been the case, however. In 1947, for example, 60% of Michigan's energy supply came from coal brought into the state. In the face of dwindling oil and gas reserves, it may be highly desirable to place a greater reliance on coal as a fuel resource. What would be the effect of basing the total energy needs of Michigan on coal?

Since Michigan now uses 35 million tons of coal, converting totally to this fuel resource would mean importing around 85 million tons of coal annually. By 1985 this annual figure could increase to 150 million tons. If it were all handled by rail transportation, there would be over 15,000 unit train trips to Michigan per year. This is nearly 45 unit trains per day, every day of the year. Current rail facilities would need to be upgraded and expanded.

Detroit Edison is currently planning a coal dock facility to handle 20 million tons per year, barely 13% of the 1985 need, at a cost of $45 million. Simple extrapolation would predict a requirement of at least half a billion dollars for docking and handling facilities alone.

Converting to a coal economy would be a tremendous undertaking. Thus, it must be carefully studied for all implications, economic, social and environmental.

Financing a conversion to a total coal economy could require from $1 billion to $1.5 billion in annual expenditures. The cost of new plants is going up drastically with a fivefold increase possible by 1990. Higher financing costs (from 4% in 1965 to 10% in 1976) have hampered efforts to raise capital and have, in part, been responsible for the erosion of the financial stability of many utilities and some industries.

Numerous jobs would be created by a coal conversion economy, but some jobs would be lost. How much labor will be displaced and how many jobs would survive the conversion?

Land dumping of the solid waste (ash) would require 600 acres covered by a layer 20 feet deep for each year of operation. Over 1.3 million tons of sulfur may have to be recovered each year to protect our clean air.

These facts dramatically reveal the need for a detailed investiagtion into the possibility of coal conversion. It is necessary to study in detail the potential risks and limitations. The acquisition of coal for use in Michigan may entail limitations resulting from:

- loss of supply from strip-mining legislation
- mine closures caused by OSHA restrictions
- labor dissatisfaction and strikes
- shortage of hopper cars

- elimination of railroad track networks
- excessive sulfur levels from coal cumbustion
- increase of coal prices

In addition, a coal-energy economy has its risks. We do not know all effects of SO_2 on health and well-being. At the present time, over 50% of the total man-created SO_2 emission comes from coal combustion. Conversion to a coal economy could greatly increase acid sulfate aerosols. A 100% increase in coal-fired electric generation may lead to as much as a 40% increase in ambient sulfate levels. Also, coal-fired plants have an estimated health risk from routine emissions of from 20 to 18,000 times greater than that associated with nuclear plants. In coal mining the excess death rate is at least 10 times higher than in uranium mining, although this differential would presumably decrease with a greater emphasis on strip mining of coal.

Despite the enormity of the proposed conversion to coal and its attendant limitations and risks, a failure to plan for greater dependence on coal can have even more severe effects. In the face of further curtailment of gas or increased prices for oil, Michigan's economy could suffer even more. A further economic decline could produce massive layoffs, labor dislocations, and eventual economic disaster. Despite the problems, the large coal reserves in this country and the ability to convert to coal more quickly than to other energy sources make such a conversion very attractive. The conversion to a coal economy is possible in the near term, although not without difficulty; in many cases it can be a direct substitute for scarce and expensive oil and gas.

A study to determine the feasibility of converting totally to coal in Michigan is outlined. The general tasks of such a study are:

1. Select a site in the Great Lakes region with significant oil, gas and coal utilization for the study.

2. Completely characterize present energy usage in the area. Determine amount, source and cost of present and anticipated future gas, oil and coal usage in this area.

3. Develop future national energy scenarios reflecting possible trends in oil, gas, coal and nuclear energy developments.

4. Develop scenarios for energy usage in the area. Several scenarios are envisioned, reflecting the spectrum of possibilities from no increased coal usage to provide a baseline for comparison to a complete coal energy dependence (to a maximum extent technically feasible).

5. For each area scenario, determine potential sources and characteristics of required coal (sulfur content, etc.). Determine uses and amounts of oil and gas displaced by conversion to coal.

6. For each scenario, determine all economic implications associated with the provision and utilization of required energy, including:

    (a)   Capital investments required to transport and utilize coal.
    (b)   Employment type and location created and eliminated by conversion.
    (c)   Anticipated environmental impact of each scenario and the costs.
    (d)   Operating economics for each user for coal and for alternate source.

7. Select the most economically attractive and technically practical scenario. This will require the consideration of possible international energy trends and the use of econometric models.

8. Determine potential government, institutional, regulatory and policy restrictions on each scenario.

9. Define specific government actions required to implement the most realistic and attractive scenario and develop a plan for implementation.

Such a program for studying coal conversion possibilities would yield two major results. First, a realistic assessment of the feasibility and implications of more extensive coal utilization, reflecting the most likely future international energy scenarios, can be determined. Second, this study will yield a plan for the attainment of the most economically viable energy program for a region. A listing of possible tasks needed for performing a coal conversion study is summarized below:

1. Describe present energy systems.
2. Prepare energy demand forecasts.
3. Develop scenarios for future energy usage.
4. Study availability of coal for each scenario.
5. Determine socioeconomic implications.
6. Define and study a coalplex.
7. Study feasibility of fuel substitution.
8. Evaluate possibilities for energy conservation.
9. Determine potential for alleviation of gas-oil curtailment.
10. Study transportation facilities.
11. Study site selection and siting requirements.
12. Perform comparison studies.
13. Select most promising scenario.
14. Determine institutional, regulatory and policy barriers to implementation.

VIII

CHARACTERISTICS OF MICHIGAN
ENERGY CENTER STUDY SITES

INTRODUCTION

As part of any energy center evaluation, specific study sites were selected. In some cases this selection had already been determined by the availability of the land, as with Camp Gruber. In other cases, specific sites were selected as part of the study so that the evaluation results would be meaningful. In all the energy center studies, there was no attempt to ensure that the best possible sites were determined. Consequently, these specific sites should not be considered as recommended by the study teams or program sponsors. They were selected only as study locations to allow a specific analysis to be performed. This chapter will provide a summarization of the physical and cultural characteristics of the study site, based on existing sources of information. Much of this information will be needed for subsequent impact analyses. A comparison of the characteristics of the various center site selections will assist the evaluation of feasibility. Site characteristic summarization identifies any additional information necessary for a more complete analysis of energy centers. Only the detailed characteristics of the Michigan energy center sites will be presented in this chapter. A summary of major differences in site characteristics for the other energy centers will be presented in the final section of this chapter.

This chapter describes the physical features of each site, including the hydrology, meteorology, geology and ecology. In addition, there is a presentation of the cultural features of each site and surrounding areas, such as population, employment and labor force, regional economics, public facilities, transportation and utilities, recreation, and historic and natural landmarks. The Michigan energy center study team selected two sites for closer examination. Site 1, located near Lake Huron in the "thumb" region of Michigan, was examined in the greatest detail. Site 2, north of Muskegon and near Lake Michigan, provided information on features significantly different from Site 1.

Both Michigan sites are located within the watershed of small streams flowing into one of the Great Lakes and are also within the watershed of larger rivers. Both sites have water readily available from the Great Lakes, but might have some surface drainage problems. The geologic conditions of the sites are very similar. As with the whole state of Michigan, both sites are in areas of very low seismic activity. There is a low frequency of severe weather episodes at either site. The regional meteorological data indicate that no unusual lack of atmospheric dispersion is common at either site. Both sites have a low population density, but the Muskegon site is much closer to some major urban centers. The Muskegon site has a limited access highway through the center of the site, while the "thumb" site has no existing or projected major arterials. The material for this chapter is largely taken from the Michigan study.[1]

THE HARBOR BEACH, MICHIGAN, ENERGY CENTER STUDY SITE

The easternmost site, referred to as Site 1 by the study team, is located in the "thumb" region of Michigan, near the town of Harbor Beach as shown in Figure 17, and in more detail in Figure 18. The site contains approximately 27 mi^2 and is located within Huron County. The site is a square, five miles on a side, with a corridor extending from the eastern edge of the site to Lake Huron, a distance of two miles.

The site is 100 miles north-northeast of the center of Detroit, 75 miles northeast of Flint, 60 miles east-northeast of Saginaw and 2 miles west of Harbor Beach. The approximate geographical coordinates of the site center are 43°51′45″ north latitude and 82°45′14″ west longitude.

The Harbor Beach site is a glacial lake bed and is, therefore, extremely flat. There are no large streams or lakes within the site. The soil type is humic gley. Approximately 85% of the total area of the site is agricultural land. The remaining area is mostly small scattered woodlots. An infrared aerial photograph is given in Figure 19, which illustrates the land cover patterns. The major crops are corn, soybeans, alfalfa, sugar beets and some grains. Figures 20 and 21 show the general nature of the site area. Figure 22 is a photograph of the shoreline within the study-site corridor. It is apparent that present shoreline development or use in this area is very limited.

Figure 17. Study site 1 location.

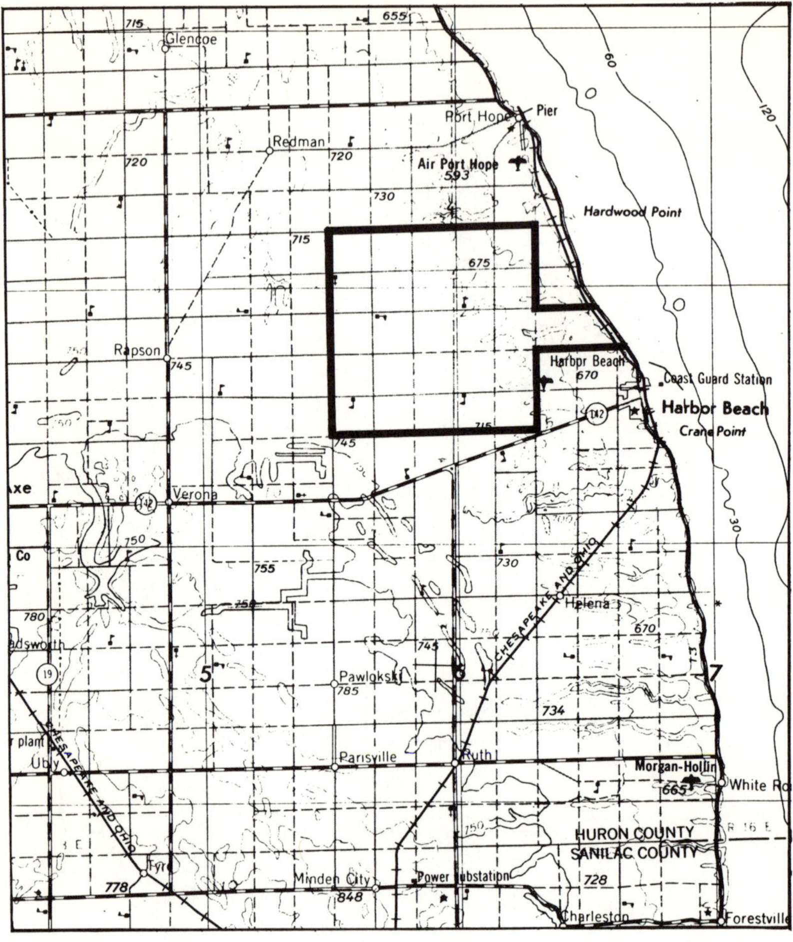

Figure 18. Regional location of study site 1.

Hydrology

As mentioned previously, an energy center will require substantial amounts of water. This water can be obtained directly from Lake Huron for either once-through cooling or as makeup water for a closed-cycle system. There are no streams in the area with sufficient flows to provide the needed supply.

Most existing water supplies in the thumb region are obtained from wells extending into either glacial drift or underlying bedrock. A few

Figure 19.   Aerial photograph of study site 1.

Figure 20. Agriculture, study site 1.

Figure 21. Agriculture, study site 1.

Figure 22. Lake Huron shore, study site 1.

towns in Huron County obtain their water from Lake Huron. The existing
population is already putting pressure on the groundwater supplies. An
energy center and new town would not be able to use any groundwater
but must obtain water directly from the lake.

The Harbor Beach site is drained by two major rivers, the Cass and the
Black, along with two minor rivers, the Pigeon and the Willow. There are
also numerous small creeks, which drain directly into Lake Huron. It is
expected that drainage from the Harbor Beach site will flow into the east
branch of the Willow River and the several creeks that flow into Lake
Huron. The site is flat and slopes gradually to Lake Huron, which allows
the runoff to occur as several small creeks rather than a few large flows.

Groundwater found in the site area comes from glacial and bedrock
aquifers located below the glacial deposits. The wells in the glacial drift
in and around the site area will provide low productivity, with yields
generally less than 10 gpm. Since groundwater and surface supplies in the
site area are quite limited, increased water demands of a center will require service from other sources.

Because of the size of the center, its many impervious surfaces and the
low infiltration rate of the site area soils, the design and layout of a
center would have to accommodate an occasional large runoff. There is
the danger that runoff from the center could produce short-term localized
flooding as well as washing the contaminants from the center into the lake.

With proper design, it may be possible to utilize a portion of the runoff as makeup water.

Lake Huron has a water surface area of 23,000 mi^2 and a drainage basin area of 49,600 mi^2. The total population of the basin was around 12 million in 1970. Six industries use the lake for waste assimilation, including three that discharge cooling water only. Four municipalities use the lake for waste assimilation. Water quality is excellent for most of the lake, although it is somewhat poorer in Saginaw Bay.

Some nearshore areas with harbors (such as Harbor Beach) and tributaries have a somewhat lower water quality than that of the lake proper. Decreased light penetration, increased nitrogen and phosphorus concentration, and general algal population increases have occurred at Harbor Beach, probably as a result of agricultural fertilizer runoff.

Meteorology

The meteorological data for the Michigan energy center was obtained from those permanent met stations nearest the sites, since the scope and limitations of the study precluded the collection of onsite data. It was felt that the nearby data would be sufficient for a preliminary evaluation. Meteorological information was obtained from stations at Flint, Saginaw and Bay City, all within 75 mi of the site. Most of the regional climatic data were obtained from Flint, which is the closest First Order Meteorological Station. Although Flint weather is not as readily influenced by the lake, it was fairly representative of the entire region. The prevailing winds are from the southwest, toward the lake, thus limiting the effects of winds from Lake Huron on the study site.

The Harbor Beach site is in the East Central Lower Climatic District of Michigan, and the area is characterized by generally flat farmland. The climate of the region is determined by frequent passages of migratory cyclones and anticyclones moving generally from west to east, and by the influences of Saginaw Bay and Lake Huron. All migratory cyclones, whether moving across the Rocky Mountains or originating in Texas, Colorado or western Canada, have a tendency to move toward the Great Lakes. The cyclone tracks tend to converge near the study site and cause more variable wind speeds and directions than other regions of the country. Except for the occurrence of severe weather, this variability would be a plus for an energy center since it would tend to cause dispersal of air pollution.

Migratory cyclones, in winter and early spring, and anticyclones in summer, generally cause frequent changes in the air mass so that the region seldom experiences prolonged periods of either hot, humid weather in summer or extreme cold in the winter.

The influence of Lake Huron on the climate at Harbor Beach is naturally strongest when there are easterly winds. In winter, this produces more snow showers but usually milder temperatures. Temperature data available show the following extremes: a high of 105°F on July 10, 1936, and a low of 24°F below zero on February 5, 1918; the warmest monthly mean temperature, 75.5°F was recorded July 1918, while the coldest month was January 1912 with 8°F.

Precipitation is well-distributed throughout the year. The crop season, May to October, receives an average of over 18 inches, or 56% of the average annual total. June is usually the wettest month, while February is the driest month. Evaporation during the crop season averages about 24 in./yr at Harbor Beach. Soil moisture replenishment during the fall and winter months is important for the success of agriculture. Extremely severe drought conditions will occur only about 3% of the time. Precipitation conditions are generally favorable for agriculture in this region and are a further indication of the desirability of locating an energy center with agricultural applications in such a region.

The average annual snowfall at Harbor Beach is 72 in., greater than most other points in southeast lower Michigan, and could cause special concern for the reliability of power transmission from this area. The average date of the last freezing temperature in the spring is May 8, while the average date of the first freezing temperature in the fall is October 13. The average growing season lasts 158 days. The use of waste heat for soil warming would extend this growing season but estimates of the extension are not available at this time.

Of particular importance for the consideration of district heating is the degree day data as an index of the heating requirements for buildings in the region. Degree days are obtained by subtracting the daily mean temperature from 65°F. Above this temperature, the need for heat is considered slight or none. At Harbor Beach, the average degree days for October is 425, while for November it is 789, indicating that nearly twice as much fuel will be required for heat in November as in October.

In this region of Michigan, there is an average of 10 to 15 days per year with dense fog, usually occurring during the winter. Although this is a small fog occurrence compared with many other areas, the number of fog days will most surely increase because of the effect of the cooling towers and waste heat discharge from an energy center. The magnitude of the increase in the number of fog days is difficult to estimate and probably cannot be determined too closely. These fog-day estimates were obtained from Flint data; it is likely that the Harbor Beach site would have more fog because it is closer to the lake and is thus subject to the formation of a lake breeze fog. This occurs when warm, moist air from the land is transported over the cold lake; if the winds are light, a dense surface fog may develop over the lake. This fog may then be carried over the land by a lake breeze during the day and recede at night during a land-breeze flow. Although these fogs rarely penetrate far inland, they can still have an effect at a shoreline energy center and, conversely, the center can magnify the effect.

The distribution of wind direction and speed is a very important factor in considering the transport, dispersion, diffusion and deposition of the potential air effluents emanating from energy-generating facilities, whether coal-fired or nuclear. Near Lake Huron, the development of local lake and land breezes in the summer during light winds and clear or partly cloudy skies may be of primary importance because of the reduced dilution of effluents from the center. It would be ideal if no atmospheric contaminants were emitted from energy-generating facilities, but this is not likely. The ideal situation is when these effluents disperse quickly and widely over noninhabited areas such as one of the Great Lakes. An energy center located on the western edge of a large lake with prevailing westerly winds, such as would occur at Harbor Beach, possesses the best possible conditions. Only the lake breezes flowing onshore, as sometimes occurs in summer, detracts from the ideal. A lake breeze is usually preceded by the passage of a lake-breeze front, which causes the following simultaneous changes: (1) a decrease or leveling off of temperature; (2) an increase in relative humidity; (3) a shift in wind direction; and (4) an increase in wind speed. Because the lake air is cooler than the land air, the breeze front tends to displace the land air as it moves inland. Vigorous mixing takes place where the two air fronts meet. However, the opposite directions of the breeze fronts tend to cause air particulates to remain localized rather than disperse over the lake where natural cleansing mechanisms can operate.

Atmospheric dilution is directly proportional to the wind speed (other factors remaining constant). Wind direction persistence is also very important when considering potential effects from a contaminant release. Wind direction persistence is defined in terms of a continuous flow from a given direction. Higher winds but lower wind persistence would be beneficial in reducing the effects of air pollution. Atmospheric stability is a measure of the degree of air turbulence. Stable conditions, or low wind turbulence, result in suppressed contaminant diffusion, while a high degree of wind turbulence, instability, is associated with enhanced diffusion conditions. Absolute measures of atmospheric stability are difficult to come by; only relative measures and direct comparisons with other sites are meaningful.

The rate of occurrence of severe weather (strong winds, tornadoes, hail and large short-term rainfall rates associated with severe thunderstorms) has a direct bearing on the reliability of power from electric power plants. It can be even more important at an energy center because of the possibility that a common mode failure could cause an outage of a greater proportion of a region's generating capacity. The mean annual occurrence of thunderstorms at Harbor Beach is 31, with June having the most (seven) followed by July and August (six). From 1955 to 1967, the average number of tornadoes reported for Michigan was 15. This is only 2% of the total observed in the U.S. This low frequency of tornadoes occurring in Michigan, which is a plus factor for siting energy centers, may result from the modifying influence of the Great Lakes. Since 1900, only one tornado is known to have touched down in Huron County. The probability estimate of a tornado striking any point in the Michigan energy center site area is about 1.2 x 10^{-4} and the recurrence interval is one in 2,000 years.

A total of 340 occurrences of wind gusts, 50 knots or greater, was recorded in Michigan for the years 1955 to 1967. In the same time period there were 100 occurrences of gusts of 65 knots and greater. Estimates indicate that an extreme wind speed of 127 mph at 30 feet can be expected to occur once in 100 years near Flint.

Meteorological records for Harbor Beach show that the greatest monthly precipitation total occurred in June 1967, with 8.66 in. November 1904, was the driest month on record with only 0.09 in. of precipitation observed. The greatest 24-hr total, at 4.00 in., fell on June 17, 1935. The heaviest single-day snowfall was 18.0 in. recorded for January 27, 1967, while the greatest recorded snowdepth was 28 in, January 13, 1969. Seasonal snowfall has varied from 141 in. in 1966-67 to only 14 in. in 1932-33. The entire thumb area of Michigan has an average of two hailstorms per year, or between 50 and 60 days with hail in a 20-yr period. Records for the Tri-City Airport, near Bay City, show the following number of hourly occurrences for freezing rain or freezing drizzle:

Year:	1949	1950	1951	1952	1953	1954
Hours:	48	48	42	44	33	62

The probable occurrence of severe weather episodes will provide estimates of how likely and how frequent the danger of power service disruption will be. But meteorology of an area is also important for determining the air pollution potential. Periods of high air pollution potential are usually related to a stagnating anticyclone with an average wind speed of 9 mph or less, no precipitation and a shallow mixing depth.

At Harbor Beach, the greatest air pollution potential occurs during the months of August, September and October. There were about 14 anti-cyclone stagnation episodes of 4-day duration reported in the Harbor Beach region from 1936 through 1965. The frequency of low-level inversions or isothermal layers below 500 ft is approximately 30% of the total hours on an annual basis in the Harbor Beach area. Most inversions occur at night.

The average maximum mixing depth is another restriction on the atmospheric dilution capability. The mixing depth is the thickness of the atmospheric layer, measured from the surface, in which convective over-turning takes place. This overturning occurs when daytime heating of the surface occurs and the warmer surface air exchanges places with the higher cooler air. The mixing depth is usually shallowest during the early morning hours just after sunrise, when the nocturnal inversion is being changed by solar heating at the surface. Mixing is at its greatest during the latter part of the afternoon between 3:00 p.m. and 4:00 p.m., when the maximum surface temperature of the day is reached. For the Flint region, the highest average morning value of the mixing depth was 939 m, while the lowest average morning depth was 280 m. The maximum mean depth in the afternoon, nearly 1,700 m, was in the summer.

Geology

The Michigan energy center site near Harbor Beach is an extremely flat former lake plain that slopes gently towards Lake Huron. Contours within the site are usually parallel to the lakeshore. The total relief within the entire main site is only 100 ft from corner to corner. There is an additional 50-ft relief in the corridor extending from the main site to the lake. The entire site is covered by lake bed sediments about 50 ft thick. The soils in the area were developed under poor drainage and consist mostly of silts and clays. As a result of glacial scour, the bedrock surface has a smooth gentle surface. A layer of Marshall sandstone underlies the lake bed deposits. The study site lies along the eastern rim of the Michigan basin. The deepest part of this geologic basin is in the center of the lower peninsula.

The region around the Michigan site is one of very low seismic activity. Some of the tectonic features appearing in the Michigan basin seem to be associated with gradual subsidence. The subsidence seems to have ceased prior to Mesozoic time. Some regional uplift of rebound has occurred since the last glacier, but this has not caused faulting or signifi-cant changes in the land surface. The nearest known major faults are in southern Ontario, Canada, some 75 mi from the site, and there are no

known active faults within 200 mi of Harbor Beach. There are no
natural caverns or extensive solution cavities known to occur in the
region. There is no physical evidence of fissuring, landsliding or caving to
indicate that there have been past earthquakes, nor should they have
been expected to occur in view of the low intensities experienced in the
past in the region. The entire lower peninsula of Michigan has a history
of low seismic activity. The maximum known intensity for any earthquake in the southern peninsula is six on the modified Mercalli scale. The
largest distant earthquake felt in Michigan was the 1812 earthquake in
New Madrid, Missouri, at a distance of 500 to 600 mi southwest. Table
XXV shows five recorded earthquakes that occurred in the Saginaw Bay
area. There is no evidence of any earthquake-related damage from these
shocks. A positive factor for siting energy centers, especially those with
nuclear units, is the indication from available geologic and seismic information that no major earthquake of an intensity greater than eight would
occur in the lower peninsula of Michigan.

Table XXV. Earthquake Epicenters–Saginaw Bay Area

Date	Maximum Intensity	Location	Felt Area (mi^2)
6 February 1872	V	Winona, Michigan	Local
17 August 1877	IV-V	Southeast Michigan near Detroit	200
16 March 1922	III	Port Huron, Michigan	Local
20 November 1930	III	Owbow, Michigan	Local
13 March 1938	II	Detroit River, Michigan	Local

Favorable foundation conditions are provided at Harbor Beach in the
form of good bedrock a short distance below the surface. The presence
of impermeable soils high in clay content make the area suitable for large
cooling-pond structures. The flatness also makes the area satisfactory for
the construction of surface ponds.

Ecology Near Harbor Beach

The construction of an enterprise as large as an energy center is going
to disrupt the ecology not only at the site but in the surrounding areas.
It is important to determine the characteristics of the current ecological
system to determine whether any species would be endangered and whether
it is possible to mostly relocate the flora and fauna. Huron County, the
site of Michigan energy study, is a predominantly agricultural region. The
Harbor Beach proposed site area is typical: the lands are devoted to

agriculture and include numerous woodlots. Major crops are winter wheat, dry beans, oats, corn and sugar beets. Wildlife in the area will feed on the grain crops and the area was once a prime pheasant hunting area. Clean farming practices have impinged on pheasant nesting cover and reduced their food supply.

The construction of deep drainage ditches in the area have developed unique habitats that offer food and shelter for numerous wildlife species. The combination of terrestrial and aquatic habitats provides a high diversity of wildlife.

The thick undergrowth of the ditch communities provides excellent shelter for pheasants, rabbits, other small mammals and occasional predators (red fox, hawks). These species feed on the fruits, foliage and barks of ash, dogwood, elm, sumac and wild grape, while the red fox, great horned owl and various species of hawks prey on rabbits, pheasants and small mammals.

"Aquatic vegetation (algae, duckweed, arrowhead, cattail) in the ditches provides food for ducks and muskrats. Aquatic vegetation during the summer months is always quite luxuriant due to the fertile runoff from surrounding fields. Ditch banks are used by muskrats as burrows. Resident ducks that utilize these communities for nesting sites are potential prey for muskrats. A diverse population of songbirds feeds on insects, seeds and fruits and utilizes the thick undergrowth to build their nests.

"The ultimate successional status of the drainage ditches is an overstory of ash, cottonwood and willow with an understory of sumac, dogwood, wild grape and various shrubs and herbaceous vegetation. Succession is kept in check by clearing fields of encroaching vegetation and by periodic dredging of the ditches."[37]

The woodlots offer nesting sites for raccoons, squirrels, wood ducks, owls, hawks and some songbirds. The fruits, seeds and bark of the many plant species provide pheasants, rabbits, squirrels, raccoons, opossums, wood ducks, other small mammals and occasional deer with an abundant food supply and cover. Predators include the red fox, the great horned owl, various hawks and man. Table XXVI lists some of the wildlife found in Huron County in 1970.

The woodlots are relatively undisturbed and should have a great diversity of insect fauna. Spiders and insects associated with the woodlots are typical of other undisturbed areas in southern Michigan. The drainage ditches, however, are relatively stagnant and turbid, and may show evidence of pollution-tolerant forms, such as rat-tailed maggots, blowflies, fleshflies and houseflies.[37]

Table XXVII is a partial list of reptiles and amphibians typical of the Harbor Beach site.

Table XXVI. Status of Wildlife as of 1970, Huron County[44]

Class and Species	Density	Trend	Notes
Big Game			
White-tailed deer	Medium	Stable	
Turkey	Low	Stable	
Waterfowl			
Ducks	High	Stable	
Geese	High	Stable	
Small Game			
Cottontail rabbit	Low	Stable	
Ring-necked pheasant	Medium	Stable	
Ruffed grouse	Medium	Stable	
Gray squirrel	Low	Stable	
Fox squirrel	Medium	Stable	
Woodcock	Medium	Increasing	
Mourning dove	High	Stable	
Snowshoe hare	Low	Stable	
Bobwhite quail	—	—	
Furbearers			
Muskrat	High	Stable	
Mink	High	Stable	
Beaver	Medium	Stable	
Weasel	Medium	Stable	
Raccoon	High	Stable	
Skunk	Medium	Stable	
Opossum	Medium	Stable	
Badger	Low	Stable	
Otter	Low	Stable	
Nongame			
Woodchuck	Medium	Stable	
Red fox	High	Stable	
Gray fox			Occasional
Crow	Medium	Stable	
Red squirrel	Medium	Stable	
Raptors	Medium	Stable	
Porcupine	Low	Stable	

The Huron County shoreline and the Saginaw Bay shoreline are characterized as wetlands. The most northern part of Huron County is composed of sandy beaches with low dunes and bluffs. The eastern shoreline of the County is exposed bedrock with very rocky shorelands especially at the point where the energy center corridor site has been suggested. In past years there have been major changes in Lake Huron's fish resources, including sea lamprey predation, an increase in alewives and fishing exploitation. Some changes in the physical and chemical environment

**Table XXVII.   Partial List of Reptiles and Amphibians
Typical of the Harbor Beach Site**

Reptiles	Amphibia
Snapping turtle	American toad
Painted turtle	Spring peeper
Blandings turtle	Tree frog
Softshell turtle	Cricket frog
Red-bellied snake	Pickeral frog
Brown snake	Chorus frog
Water snake	Leopard frog
Garter snake	Green frog
Hognose snake	Wood frog
Ringneck snake	Bull frog
Green snake	Mud puppy
Blue racer	Blue-spotted salamander
Rat snake	Red-backed salamander

have been important; water pollution from agricultural runoff has
diminished fishing quality in the thumb area and erosion and siltation
from agriculture and construction projects have contributed to a degradation in water quality. The main body of Lake Huron, however, has
shown only slight changes; increases in total dissolved chlorides and sul-
fates are evident. The construction and operation of a large-scale energy
center in this area can be expected to result in further problems in
reducing environmental quality of the region unless great care and planning are performed. The expected impacts on the ecology from an
energy center are described in detail in later chapters.

Population

The towns and villages of Huron County are small and serve as local
market centers. The major urban areas of the region, Bay City, Saginaw,
Flint and Detroit, are all more than 60 mi from the Michigan study site.
Tables XXVIII, XXIX and XXX give population statistics for several
counties in the thumb region, including Huron County. These statistics
reflect the predominantly agricultural aspect of the region, especially by
the low population densities. Of particular importance is the suggestion
of population decreases in Huron and Sanilac Counties. Within the study
site itself, there are about 153 residences or a population density of
approximately 27 persons per square mile. The decrease in population in
Huron County appears to be a result of younger families seeking better
employment opportunities and the increased mechanization of the farms.

Table XXVIII.   Components of Population Change by County:
Census, 1950 to 1970[38]

| County | 1950 to 1960 | | | 1960 to 1970 | | |
	Population Change	Natural Increase	Net Migration	Population Change	Natural Increase	Net Migration
Huron	857	5,830	-4,973	77	3,615	-3,538
Lapeer	6,132	5,672	460	10,391	5,837	4,554
St. Clair	15,602	15,018	584	12,974	11,529	1,445
Sanilac	1,477	4,054	-2,577	2,575	2,736	-161
Tuscola	5,047	6,020	-973	5,298	4,915	383

Table XXIX.   Population of Selected Michigan Counties:
Census 1940, 1950, 1960, 1970; Projected 1980, 1990[38]

County	1940	1950	1960	1970	1980	1990
Huron	32,584	33,149	34,006	34,083	31,765	29,110
Lapeer	32,116	35,794	41,926	52,361	62,936	74,998
St. Clair	76,222	91,599	107,201	120,175	130,862	141,734
Sanilac	30,114	30,837	32,314	35,181	36,554	38,526
Tuscola	35,694	38,258	43,305	48,603	53,831	60,141

Table XXX.   Distribution of Population by County: 1970[38]

| County | Population Density/mi^2 | Urban Population | | Rural Population | |
		Total	Percentage	Total	Percentage
Huron	42	2,999	9	31,084	91
Lapeer	80	6,270	12	46,047	88
St. Clair	164	55,320	46	64,855	54
Sanilac	36	0	0	34,889	100
Tuscola	60	6,503	13	42,100	87

Labor Force and Regional Economics

The 1970 census indicated a labor force of 10,821 in Huron County,
compared with 17,090 in Lapeer County and 3,252,830 for the entire
state. Along with the small labor force, the region has a serious unemployment rate. The low existing labor force within this area would
certainly not be sufficient to support the construction or operation of an
energy center.

The thumb area, with its chronic high unemployment and its small
manufacturing base, can be characterized as economically depressed.
Most of the economic base for the area is centered on agricultural
production and the supporting retail and service activities. Tourism is

not an important aspect of the regional economy as in other areas of
Michigan. Another indicator of the economic situation in Huron County
is family income. Over 12% of the thumb-area households had an income
under $3,000, compared to a statewide average of 7.5%. The number of
families with incomes of $10,000 or greater is only 34.2% in Huron
County, compared with 57.2% statewide. Low incomes mean less money
to purchase goods and services which generate more employment and
increase per family income.

Manufacturing has a broad base but has few companies with a small
number of employees. The retail establishments are small but adequate
for existing populations. Major chain stores are not available except very
near the larger cities. Average weekly earnings by county are shown in
Table XXXI. It is evident that the wage rates within the thumb counties
are considerably lower than other areas.

Table XXXI.   Average Weekly Earnings by County, 1972[38]

County	Average Weekly Earnings ($)
Genesee (Flint)	207
Huron	131
Lapeer	149
Saginaw (Saginaw)	194
Sanilac	137
Tuscola	142
Wayne (Detroit)	206
State of Michigan	184

Public Facilities and Institutions

There are 14 schools within 15 miles of the Michigan study site. The
Sigel Number 4 school is on the study site itself but has an enrollment
of only 27 students, and most likely will be closed soon. The total
student enrollment for Huron County in 1975 was 8,543 and has shown
a decrease over the past several years. There are no facilities for educa-
tion beyond high school in the county. Most schools are currently
operating near maximum capacity and could not support a substantial
increase of new students resulting from an energy center.

Huron County has the following medical facilities[41]:

Medical Facilities in Huron County[38]

3 hospitals, total of 146 beds	204 registered nurses
3 nursing homes, total of 137 beds	15 dentists
26 physicians	2 dental hygienists

The city of Harbor Beach has a hospital with 29 beds, 5 doctors and 3 dentists. These facilities are the minimum needed for the existing populace and could not provide adequate services for any large influx of population from an energy center or new town of 200,000 population.

Municipal Services

The present sewage treatment plant at Harbor Beach has a capacity of 350,000 gpd. This facility currently services all the households within the town. The plant is undergoing expansion to an anticipated capacity of 1 mgd. The present water system has a capacity of 2 mgd and obtains its water from Lake Huron. Other services in Harbor Beach include a small weekly newspaper, library, radio station, volunteer fire department, a small police force, and municipal garbage and refuse pickup.

Transportation and Utilities

The present transportation facilities around the study site are adequate for the low population and industrial activity in the area, but would be totally inadequate for the expansion of an energy center. However, the low relief and open expanse of this area would allow easy expansion of the transportation facilities of the area to meet the increased demands that would occur with an energy center.

Current highways near the site are U.S. Highway 25, which follows the shoreline of the entire thumb and passes east of the site through the extension to Lake Huron, and State Highway 142, which runs west from Harbor Beach along the southeast corner of the study site. There are currently no plans to extend major highways into the area. There are no airports with commercial carrier service in the thumb region at this time. Rail service at Harbor Beach is limited to one line serviced by the Chesapeake and Ohio Railroad with one freight train every day except Sunday. The train, of 7 to 20 cars, is used for grain, peat and coal for the power plant at Harbor Beach.

Of interest to energy centers is the harbor for Great Lakes shipping at Harbor Beach. This facility has a large breakwater and a narrow channel with a maximum depth of 20 ft and leads to the Detroit Edison Power Plant. The power plant obtains most of its coal by boat. The harbor can handle 600-ft boats in the channel to the power plant; however, due to the shallow depth of the channel, these boats must not be fully loaded to avoid grounding. The harbor was manmade and could be dredged to handle larger boats. The harbor will freeze during some winters and industries relying on the harbor must stockpile supplies.

The only electrical generating facility in this area is the Detroit Edison Plant at Harbor Beach which has a single coal-burning steam boiler with a 122-MWe capacity. No petroleum pipelines extend to the thumb area and there is one natural gas pipeline which extends to Harbor Beach.

Recreation Facilities and Historic and Natural Landmarks

Recreational activities have not been a dominant feature of the thumb area because of the predominant agricultural activity, few forests and the lack of unique natural features in the region. Within Huron County there are two state parks, eight county parks and two state game areas, shown in Figure 23. These parks are all located along the shore of either Lake Huron or Saginaw Bay and lie in the northern portion of the county.

The National Register of Historic Places[39] lists four sites of historical significance in Huron County, but none are within the study site proper. There are no natural landmarks designated in Huron County.

Land Use and Availability

The study site at Harbor Beach contains primarily privately owned lands, about 85% in agricultural use and the rest in small scattered hardwood lots. The availability of land for use as an energy center is governed by two factors: (1) its market availability and price, and (2) the dedication of the land by one or more levels of government to a specific use or condition. In Michigan, there is a special situation with respect to the availability of agricultural and open-space land which will directly affect the location of an energy center in the state. In 1974 Michigan passed the Michigan Farmlands and Open Space Preservation Act, which provides for the formation of a contract between the land owners and the state, whereby the owner agrees to maintain his land in its current use for a set number of years in return for property tax benefits. The intent of the Act is to protect open land from the usual "potential use" taxation policies which often force selling of the land to developers. The implications of the Act for the development of large energy centers are obvious. Either the government's intention for land use must be redefined or certain "lightly populated, agricultural lands" would be ineligible candidates for energy center sites. Dispersed siting would seem to present less of an obstacle than a large 25-square-mile plot, inasmuch as smaller blocks of land not under a preservation agreement are more easily found.

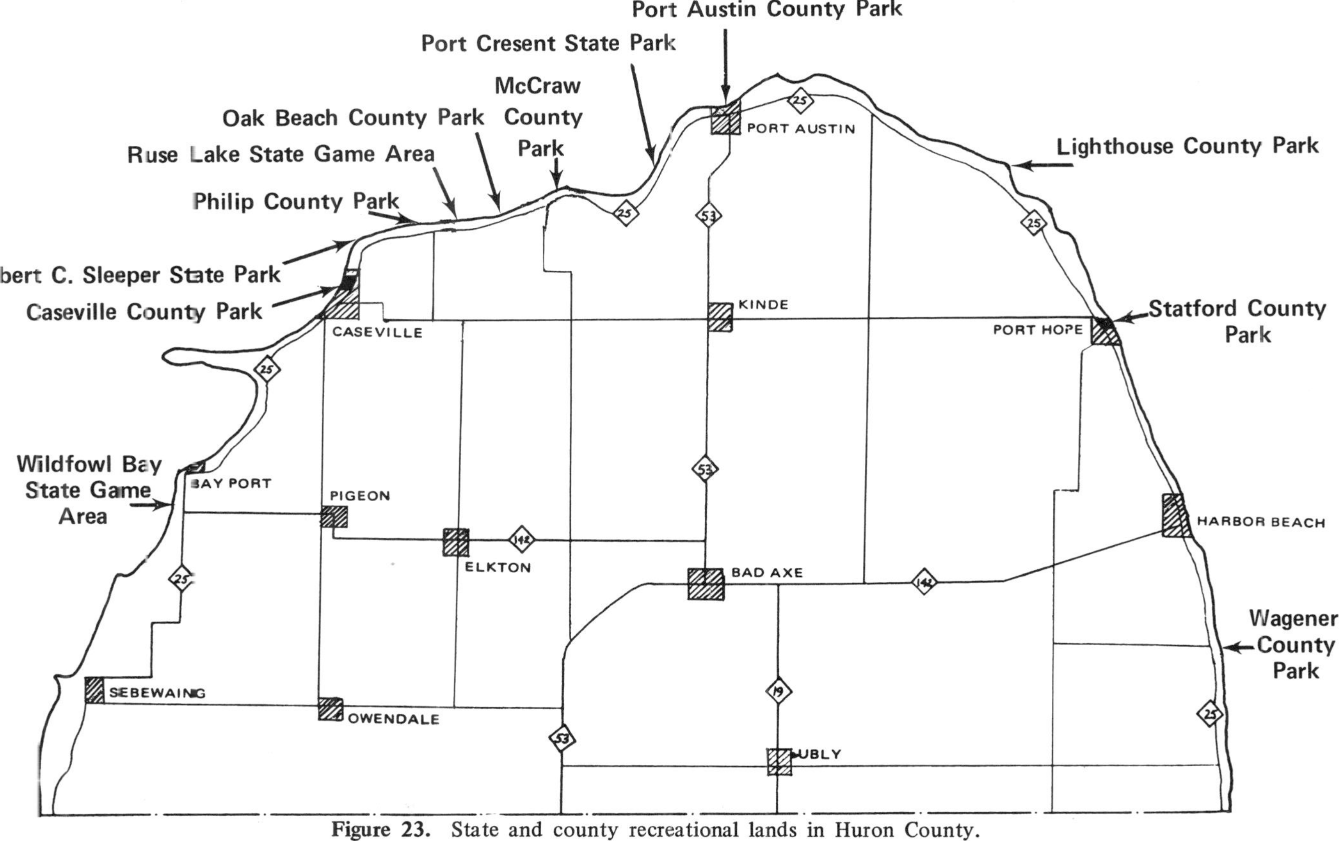

Figure 23.   State and county recreational lands in Huron County.

THE LOCATION AND CHARACTERISTICS OF
THE WESTERN MICHIGAN STUDY SITE

The location of the second Michigan study site is shown in Figure 24, with more detailed information presented in a topographic map of

Figure 24. Location of energy center study site 2.

Figure 25 and an aerial photograph in Figure 26. This site is located partly in the southern portion of Oceana County and a northern portion of Muskegon County on the western edge of the lower peninsula. The 27-mi^2 site is within 0.5 mi of Lake Michigan, about 50 mi northwest of Grand Rapids, 20 mi north of Muskegon and within 1 mi of the industrial area of Whitehall-Montague. U.S. Highway 31 runs north and south through its center. The geographical coordinates of the center of the site are 43°27'52" north latitude and 86°21'21" west longitude.

The western Michigan site consists of gently rolling glacial outwashes, moraines and lake bed sands. There are no large streams or bodies of water within the site. The western half is predominantly agricultural with major crops of corn, alfalfa, clover and vegetables. Figure 27 shows the agricultural nature of this part of the site. The eastern half of this area is a scrub oak forest, which is a part of the Manistee National Forest (see Figure 28).

Hydrology

The groundwater supplies are similar to those at Harbor Beach. The presence of sandy soils at the Montague site implies greater infiltration and the relatively higher flow of streams and rivers indicates that this site would probably have less of a runoff problem. Although this site has a larger network of rivers and streams with greater flows than at Harbor Beach, they are not sufficient to provide any cooling or makeup water for the energy center. The water quality of the rivers and streams and of Lake Michigan near the site is usually fair to good and efforts to maintain this quality need to be enforced if energy facilities are sited here.

Meteorology and Climate

The closest meteorological station is at Muskegon, about 20 miles south and near the Lake Michigan shore. The climatological characteristics of the study site should be very similar to those at Muskegon. Like the Harbor Beach site, there are no significant topographic disruptions within the site and it is located near one of the Great Lakes. The weather, then, is subject to similar climate-moderating influences from the large body of water nearby. The effects of the lake are more dominant for the western site because of the prevailing westerly winds. Normal, mean and extreme climate conditions at Muskegon are very similar to those at Harbor Beach. The western site is slightly warmer, has more snow, and has less total precipitation. The mean annual wind speeds and wind variations are similar between the two sites and, consequently, the dispersal of airborne contaminants will be about the same. The only major

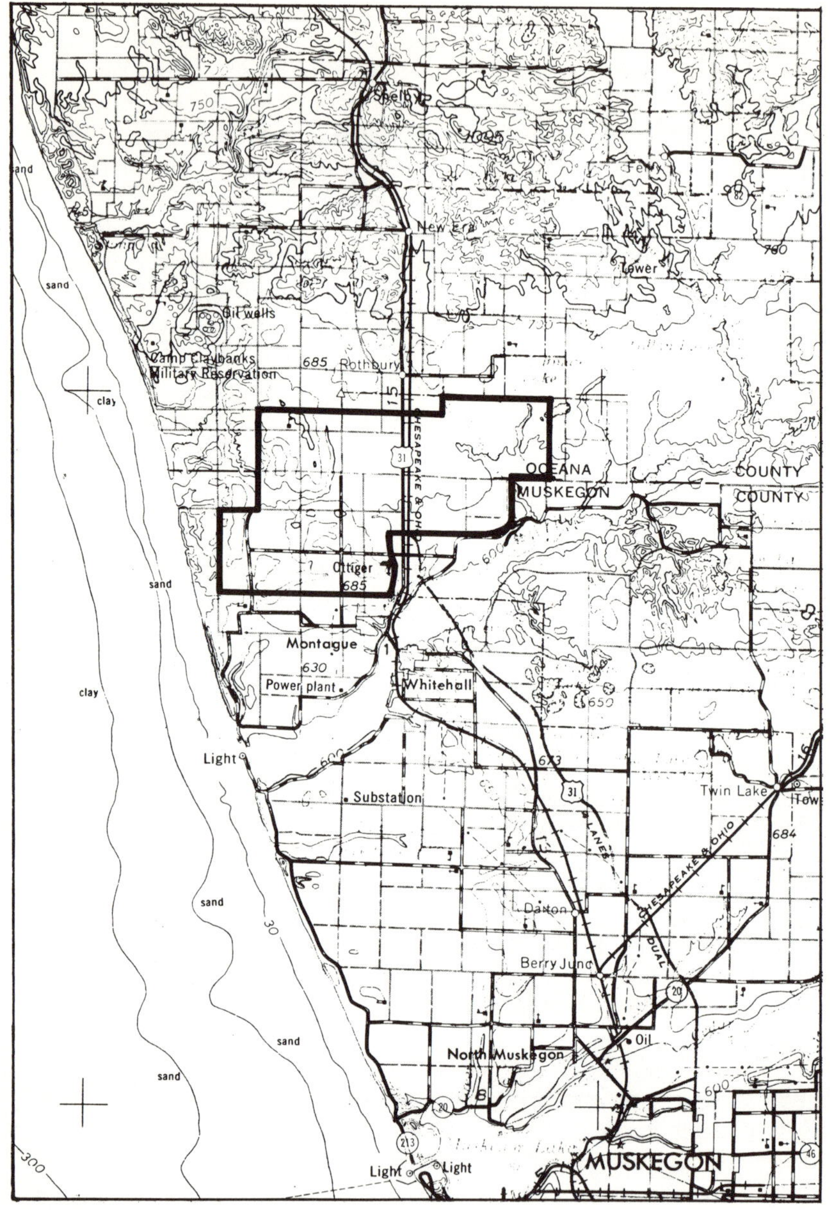

Figure 25.   Regional location of study site 2.

Figure 26.   Aerial photograph of study site 2.

Figure 27. Agriculture, study site 2.

Figure 28. Forest, study site 2.

difference is that the dispersal area for the eastern site is usually over Lake Huron, while the dispersal area for the western site is the center of the state.

The Montague study site has a higher incidence of heavy snow showers and thunderstorms resulting from moisture picked up from Lake Michigan.

Geology

The geologic conditions near the western site are similar to Harbor Beach because they lie on opposite sides of the Michigan basin. This area also has very low seismicity. The nearest known faults are in central Wisconsin and northern Illinois, at a distance of 125 mi, and are not considered active. The surface conditions are somewhat different. Glacial deposits at the Montague site consist of glacial outwash, moraines and lake bed sands. The most prominant topographic feature is the sand dunes with reliefs of 100 ft between the study site and Lake Michigan. The general soil type is podzol and brown podzolic and there are small petroleum fields and some salt, sand and gravel production in the region.

Ecology

The Montague site falls within the hemlock, white pine and northern hardwoods forest region. Sugar maple and beech are the dominant hardwood species. The southern boundary of Muskegon County is the southern limit of the oak-hickory forest zone. Historically, fires and logging have kept the forests in a state of flux. This continuous change provides a variety of cover for wildlife species such as kirtland's warbler, short-tailed grouse, white-tailed deer and ruffed grouse. Lately, fire control methods have produced a more homogeneous habitat so that wildlife management practices must be employed to maintain some species of wildlife. The recent status of wildlife in the general area of the Montague site is presented in Table XXXII.

The fish populations of Lake Michigan have undergone significant changes during the past 30 or 40 years. Originally the coldwater species predominated, including lake trout, whitefish, herring and chubs. Dramatic fluctuations in the numbers and kinds of fish have occurred because of overfishing, invasion by the sea lamprey through the St. Lawrence Seaway, increase in the number of alewife and large-scale stocking of salmonids. Herring and walleye are almost extinct and yellow perch and lake trout stocks have been reduced. Whitefish and alewife now dominate the commercial catch. The generally good water quality and alewife food supply have made Lake Michigan an excellent producer of trout and salmon. The trout fishery may grow by natural reproduction, but it appears that the

**Table XXXII.   Status of Wildlife as of 1970
Northern Lower Peninsula, Michigan[41]**

Class and Species	Density	Trend
Big Game		
White-tailed deer	Medium	Decreasing
Black bear	Low	Decreasing
Turkey	Low	Increasing
Elk	—	—
Waterfowl		
Ducks	Medium	Stable
Geese	Medium	Increasing
Small Game		
Cottontail rabbit	High	Stable
Ring-necked pheasant	Low	Stable
Ruffed grouse	High	Increasing
Gray squirrel	Medium	Increasing
Fox squirrel	Medium	Increasing
Snowshoe hare	Low	Decreasing
Woodcock	High	Increasing
Mourning dove	Medium	Stable
Bobwhite quail	Low	Stable
Furbearers		
Muskrat	Medium	Decreasing
Mink	Medium	Stable
Beaver	High	Stable
Weasel	Medium	Stable
Raccoon	High	Increasing
Otter	Low	Decreasing
Skunk	High	Increasing
Opossum	Medium	Increasing
Badger	Low	Stable
Nongame		
Woodchuck	Medium	Stable
Porcupine	Low	Decreasing
Red Fox	Medium	Stable
Bobcat	Low	Decreasing
Crow	High	Increasing
Raven	Low	Stable
Red squirrel	Medium	Increasing
Coyote	Low	Stable
Raptors	Medium	Stable
Rare (R) Endangered (E) Status Undetermined (S)[a]		
Bald eagle (E)[b]	Low	Decreasing
American osprey (S)	Low	Decreasing
No. Gr. prairie chicken (R)	Low	Decreasing
Kirtland's warbler (E)	Low	Decreasing
Eastern pigeon hawk (S)		

Table XXXII. Continued

Unusual or Unique Animals[c]		
Sandhill crane	Low	Increasing
Spruce grouse	Low	Stable
Golden eagle		
Sharp-tailed grouse	Low	Decreasing

[a]From *Rare and Endangered Fish and Wildlife of the United States,* U.S. Bureau of Sport Fisheries and Wildlife, 1968 ed. Also based on February 1972 data from the Bureau's Office of Endangered Species.

[b]For our purposes, the northern and southern subspecies of bald eagle are listed here as bald eagle, the endangered status being the important consideration.

[c]Animal species considered to be unusual or unique on a regional, state or planning subarea basis.

salmon fishery will require maintenance stocking from now on. The Montague site is situated at the center of the prime salmon fishing grounds in Lake Michigan. Both coho and chinook salmon have been successfully planted in the Muskegon and White rivers. Inland lakes and streams in the Montague study areas are important aquatic habitats for brook trout, brown trout, steelhead, northern pike, walleye, large- and smallmouth bass and assorted panfish.

Population

The primary differences between the two Michigan study sites are the large urban areas in the nearby region and the larger number of smaller towns close to the Montague site. A summary of the population distributions at and around the Montague site are presented in Tables XXXIII and XXXIV. The most significant difference between the two Michigan sites is that the higher population densities near the Montague site would be sufficient to support the construction and operating activities at the center without a large influx of labor. Center construction at this site would help the current unemployment problem the area now suffers, which has contributed to a net emigration from Muskegon County.

Labor Force and Regional Economics

The labor force in the Montague area is much larger than the Harbor Beach area; the City of Muskegon had a labor force of 79,000 as of June 1975, with an unemployment rate of 16.7% or about 13,200 persons. This is more than enough to provide the 8,300 operating personnel estimated for the energy center, although it is unlikely that the

Table XXXIII.   County Population Tables[38]
Population of Michigan Counties, Census 1940,
1950, 1960; Projected 1980 and 1990

County	1940	1950	1960	1970	1980	1990
Muskegon	94,501	121,545	149,943	157,426	158,547	158,064
Oceana	14,812	16,105	16,547	17,984	18,393	20,021

Distribution of Population by County, 1970

County	Population Density/mi^2	Urban Population Total	Urban Population %	Rural Population Total	Rural Population %
Muskegon	314	108,733	69	48,693	31
Oceana	34	0	0	17,984	100

Components of Population Change by County, Census 1950 to 1970

County	1950–1960 Population Change	1950–1960 Natural Increase	1950–1960 Net Migration	1960–1970 Population Change	1960–1970 Natural Increase	1960–1970 Net Migration
Muskegon	28,398	28,430	-32	7,483	21,185	-13,702
Oceana	442	2,193	-1,751	1,437	1,424	13

Table XXXIV.   Towns and Cities in Areas Adjoining Study Site 2

Town/City	1970 Population	Distance from Site (mi)	Direction from Site
Muskegon County			
Fruitport	1,409	26	SSW
Lakewood Club	590	8	SE
Montague	2,396	3	S
Muskegon	44,631	17	SSE
Muskegon Heights	17,304	20	S
North Muskegon	4,243	15	SSE
Norton Shores	22,271	23	S
Ravenna	1,048	28	SE
Roosevelt Park	4,176	20	SSE
Whitehall	3,017	4	S
Wolf Lake	2,258	19	SE
Oceana County			
Hart	2,139	15	N
Hesperia	877	17	ENE
New Era	466	5	N
Pentwater	993	22	N
Rothbury	394	2	N
Shelby	1,703	9	N
Walkerville	319	20	NW

many special skills necessary would be well represented in the unemployed force available. Nevertheless, by comparison, there is a much larger and more qualified labor force available in the area surrounding the Montague site than exists at the Harbor Beach location.

The regional economics of the Montague area have a greater dependence on industrial and recreational activities and less dependence on agriculture than the Harbor Beach area. This will have implications for the development of an energy center. With an already established industrial base, supportive industry, labor force and a transportation network, many external requirements for supporting the center will already exist. On the other hand, the center might have a negative impact on the recreational activity in the area. The industrial activities existing near the Montague site include pulp and paper mills, and furniture, chemical and pharmaceutical, and boat manufacturing plants located in the city of Muskegon, Whitehall and Montague. There are several agricultural products processors in northern Oceana County.

Transportation and Utilities

A major difference between the eastern and western study sites is the availability of existing transportation facilities. With its higher population, greater industrial development and recreational activities, transportation facilities are better developed for the Montague site. U.S. Highway 31, a divided four-lane limited-access highway, presently divides the study site in half. This highway offers excellent connections to the major urban centers in the state. Railroad tracks run north-south through the site also, and belong to the Chesapeake and Ohio Railroad. There is a natural gas line running through the study site and a petroleum pipeline about 17 mi south at Muskegon. Air carrier service is provided at Muskegon and a basic utility airport is located at Hart. Existing airport facilities would appear to be adequate to accommodate the demands created by an energy center.

The nearest electrical generating plant is the Cobb facility at Muskegon which is operated by the Consumers Power Company. This plant has five coal units and a total capacity of 510 MWe. Transmission lines consist of a 138-kV line running north-south through the study site and several 345-kV lines east of the site running north to the pumped storage facility near Ludington.

Recreation Areas and Historic and Natural Landmarks

There are four state parks near the site: Charles Mear and Silver Lake in Oceana County, and Muskegon and Hoffmaster in Muskegon County.

The Manistee National Forest intersects the eastern portion of the study site. The Lake Michigan shoreline, particularly the dune areas to the west, has recreational value. There are six items of historic significance near the Montague site, but none within the site proper.[39] There are no designated natural areas in either Muskegon or Oceana counties.

Land Use and Availability

The land considered for the Montague site consists of private agricultural land and forested areas, including much of the Manistee National Forest. The possibility of constructing an energy center at this site would be affected both by the Michigan Farmlands and Open Space Preservation Act of 1974, and the difficulty of use of public-dedicated lands such as a national forest. Federal- or state-dedicated lands comprise unique areas that should be preserved. Although trading of land is often done to consolidate forest boundaries and eliminate private holdings, it is not likely that such trades would occur since replacement of the land with land of equivalent value would be difficult. An energy center at the Montague site is probably not feasible under present regulations because the national forest lands would be required.

CHARACTERISTICS OF SITES FOR OTHER ENERGY CENTER STUDIES

One study which extensively investigated the characteristics of a specific site was performed by the Pacific Northwest Laboratories of Battelle Memorial Institute for the area of Camp Gruber, Oklahoma. A discussion of the characteristics of Camp Gruber which follows was derived from the Battelle study[2] and provides a comparison with the characteristics of the Michigan site at Harbor Beach.

Land

Camp Gruber is located east of Muskogee, Oklahoma (Figure 29). It consists primarily of sloping uplands and lies between the Arkansas River on the west and the Illinois River on the east. The elevation varies from 1,000 ft at a line running approximately northwest to southeast to about 500 ft at Greenleaf Creek, which runs northeast to southwest. The west central portion of the site near Braggs is former flat farmland. A 930-ac recreational lake in Greenleaf valley was created by a dam on Greenleaf Creek. The 65,000-ac former army camp is owned by the state of Oklahoma and the U.S. Government. This ownership and the fact that it is unoccupied—it is currently used for recreation, grazing, game conservation

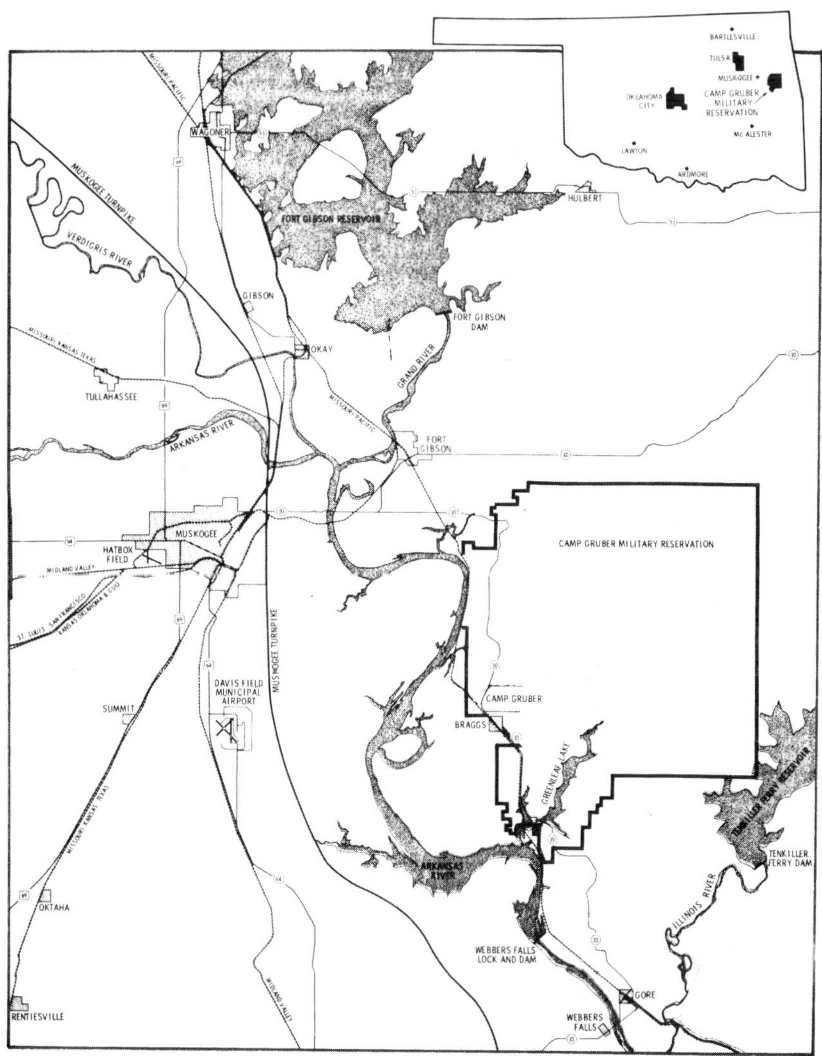

Figure 29. Map of the Camp Gruber, Oklahoma, area.[2]

and National Guard maneuvers—has suggested its use as an energy center.
Much of this land is under utilized and some has been declared surplus by
the government. Current policy encourages use of surplus federal lands
for energy facilities.

There are a few permanent facilities at Camp Gruber: one state highway,
one railroad, and several power and telephone lines. A large number of

foundations, concrete slabs, dirt roads and a few structures remain from previous army and agricultural activities. The primary recreational uses are the picnicking, camping and water sports in the area of Greenleaf Lake, and hunting on most of the camp. The recreational facilities include cottages, lodges, camps, boat ramps and playgrounds.

Geology

Camp Gruber is located in northeastern Muskogee and southwestern Cherokee Counties. The camp is on the southwest flank of the Ozark uplift and consists primarily of low mountains and hills with growths of hardwoods, cedars and scattered pines. It is drained by numerous small streams which have gravel and rock bottoms.

The primary subsurface structure of Camp Gruber consists of 10,000 ft of limestone, shale and sandstone sediments overlying a Spavinaw granite. Soils in the area are mostly of the Hector-Linker Soil Association, which are characterized as acidic and shallow, and are generally of low fertility. These soils are severely limited in use and can support only scrub oak and some grass. Such land cannot be used for productive agriculture, only for range or recreational purposes. The main geological formation covering 90% of Camp Gruber is the Atoka Formation, which consists of sandstone, shale and a few thin beds of limestone. In some areas, the Atoka Formation is several hundred feet thick with an overburden ranging from a few inches to several feet. The Atoka Formation has foundation conditions which are suitable for heavy duty construction and has a bearing capacity of from 20,000 lb/ft^2 to around 50,000 lb/ft^2.

Northeastern Oklahoma is traversed by many northeast-southwest faults and folds which can have displacements as much as 700 ft. The major fault within the Camp Gruber boundary is the Greenleaf Lake fault which runs through the eastern half of the Camp and has a displacement of from 40 to 175 ft. These faults are considered very old and inactive. The Camp Gruber area lies within a seismic risk zone in which minor damage has occurred from earthquakes. These seismic risk zones are assigned a value of from 9 to 3 in order of increasing seismic risk. The Camp Gruber area is considered relatively safe from future earthquake damage.

Geologically speaking, Camp Gruber is a favorable site for a large-scale energy center. The land is considered to be basically unproductive surplus land and therefore more readily available than land elsewhere. Camp Gruber area is not expected to contain commercial quantities of gas or oil, deposits of coal or metal ores. The area is not designated as a known geothermal resource.

Ecology

A large part of Camp Gruber is considered closed game reserves for certain species and most of the area comes under game and wildlife management. The terrain varies from rolling tall-grass prairies in the western portion to oak-hickory hills in the eastern half. Camp Gruber is part of the Ozark Mountain uplift.

The climate is mild, having an average mean temperature of 59°F. The average rainfall is about 41 in.; snowfall is about 6 in. The growing season is from 179 to 205 days and extends from mid-April to the last of October.

Along with the oak-hickory habitat, there are relatively dense stands of second-growth timber including post oak, blackjack oak, black hickory and shagbark hickory. Understory includes sumac, greenbriar, coral berry, johnson grass and others. The open fields are covered by bluestems, grama, indiangrass and cheatgrass. Along the banks of the streams and lowland areas the typical trees are cottonwood, willow, elm, ash, sycamore, red maple, silver maple and several species of oak. The bottomlands understory includes greenbriar, virginia creeper and wild grape. Lands along the Arkansas River are used as crop lands where soybeans, cotton, corn, peanuts, grain sorghums, wheat and hay are cultivated. The mild climate and high rainfall help offset the effect of poorly productive soils as long as intensive fertilization and soil management practices are used. Beef cattle are an important agricultural product and grazing rights are leased in the area.

The oak-hickory and bottomland forests and abandoned fields provide attractive habitats for wildlife. Animals prevalent in the area include waterfowl, shorebirds, marshbirds, upland game birds, songbirds, small mammals, furbearers and deer. Big game and upland game hunting is extensive; yearly, about 175,000 man-days are spent in hunting deer, quail, rabbits, other small game animals and waterfowl. A list of the threatened species of animals which may occur in the general Camp Gruber area is listed in Table XXXV.

There would be several ecological impacts in siting an energy center at Camp Gruber. Although some of the land is surplus, a major concern of the study team was the potential removal of lands from public use. Public acceptance and the various safety factors concerning hunting around energy installations is of concern. The ecology will also be impacted by the construction of fences which might disturb the migration routes of certain animals. The construction of facilities might reduce the animal population or produce deleterious effects on the interactions of some animals. Powerline construction and clearing the right-of-way may

Table XXXV.   Threatened Species of Animals at Camp Gruber

Common Name	Status
Red Wolf	Endangered
Indiana Bat	Endangered
Ferruginous Hawk	Rare
Prairie Falcon	Rare
American Peregrine Falcon	Endangered
Whooping Crane	Endangered
Greater Sandhill Crane	Rare
Southern Bald Eagle	Endangered
Eskimo Curlew	Endangered
American Osprey	Rare
Black-Capped Vireo	Rare

affect some biota.   Although there is no deleterious effect from high voltage lines known at this time, it is possible that the trend to UHV transmission lines could affect some wildlife.   Clearing and maintaining powerline corridors could be beneficial to wildlife since cleared spaces can provide the type of forage better for wildlife.   The effect of the emissions and refuse from the proposed facilities would have little or no effect on the terrestial environment if proper controls are maintained.

Hydrology

It is very desirable for an energy center to have: (1) a readily available water supply, (2) high water purity, (3) low water supply costs, and (4) large liquid waste and heat disposal capability.   The last item is critically important for large-scale steam-electric generating plants, which require large quantities of water for waste heat disposal.

At the Camp Gruber site, the major water requirement for the energy center is the cooling water used to dissipate waste heat generated during the conversion process.   In addition, smaller quantities of potable water, municipal water, are required for sanitation and other uses.   Other quantities of water would be required for any associated industries such as coal-gasification plants or petroleum refineries, or other industries that might be sited at Camp Gruber.

In the past both coal and nuclear power plants were sited near large water bodies, when possible, to provide for cooling water needs.   In these cases, once-through open-cycle cooling was used to remove waste heat from the plant condensers and to dissipate it directly into adjacent water bodies. Now, however, the Environmental Protection Agency has decreed that all power plants must incorporate the best available technology to minimize thermal discharge by 1985; this usually means closed-cycle cooling.   In closed-cycle cooling, the heat is removed from the condensers by a circulating water flow and is dissipated directly into the atmosphere by means of evaporation. This is achieved through the use of cooling towers or cooling ponds.

In some applications at an energy center, a water body must be available to receive the wastes that result from some plant operations. This water body must be of such a size and water quality that the discharge of wastes will not violate state or federal water quality standards. If the standards cannot be met using the available water body, then the use of settling ponds, brine concentrators, dry cooling towers or some other method must be considered to remove these wastes.

Approximately 1,700-1,900 ft^3 of cooling water are required to maintain an environmentally acceptable heated-water discharge temperature for a 1,100-MWe electrical nuclear plant. The Camp Gruber study team conducted an analysis to determine the maximum number of large nuclear plants or equivalents that could be located at the site. The plants were assumed to utilize cooling towers to dissipate waste heat; the blowdown from the cooling towers was assumed to be discharged without treatment into the Arkansas River. The Oklahoma water quality thermal standards were assumed to be the limiting criteria for the blowdown discharge to the Arkansas River. The critical period at Camp Gruber would be summertime, with high ambient temperature and low water flow in the adjacent water bodies. The conditions assumed at this critical time were[2]:

1.	Heat discharged per unit	1×10^8 Btu/hr
2.	Cooling water evaporation for each unit	40 cfs
3.	Arkansas River low flowrate	520 cfs
4.	Maximum ambient water temperature	86°F
5.	Maximum temperature rise of 5°F to a maximum of 93°F	
6.	Water available for cooling towers	1,000 cfs

Using these assumed conditions, Battelle estimated that eleven 1,100-MWe nuclear plants would require a total of 1,000 cfs of makeup water. This is a serious constraint on the size of any energy center that must obtain makeup water during periods of low flow from nearby rivers. However, dry cooling towers, which require no cooling water, and other technical advancements are presently under development.

The region containing Camp Gruber has four major rivers, the Arkansas, Grand, Verdigris and Illinois. Eight manmade reservoirs have been constructed on each river for flood control and hydroelectric power generation. It is possible that some water in the reservoirs could be made available for the energy center. Considering the storage capacities of the reservoirs and annual and seasonal streamflow records of the water supply to the reservoirs, it is apparent that there would not be enough water for open-cycle cooling for even a few power plants. Using the Arkansas River and a closed-cycle cooling system, Camp Gruber would be able to support the generation of only 4,000-5,000 MWe. The Arkansas River alone could not support a power plant energy center and alternative water supply sources must be considered. This will create difficulties in water allocation priorities and increased costs for water transportation.

The study team concluded that an energy center with a generating capacity limited to around 15,000 MWe of electric power appears feasible for Camp Gruber. "This mode of operation would require approximately 1,000 cubic feet per second of good quality cooling water, almost none of which is presently available. Alternatives which could supply this demand are: (1) reallocation of existing water for energy center use, (2) utilization of pumped storage at Fort Gibson or Tenkiller Ferry Reservoir, or (3) construction of new reservoirs for energy center water storage, (4) cleaning up the Arkansas River, or (5) use of brine concentrators."[2] The water supply problem at Camp Gruber seems to be the most important limiting factor to siting power generation and industrial facilities, and may be contrasted to the readily available water supply for the Michigan energy center from the Great Lakes.

A brief review of the hydrological properties of the region around Camp Gruber will help put the water supply problem in proper perspective.

Groundwater

Most wells in the area yield from a fraction of a gallon per minute to a few gallons per minute of fair- to poor-quality water. The Arkansas River valley alluvium in the Muskogee area will yield 100 to 500 gpm, although very little is near or on the Camp Gruber Reservation. Deeper aquifers at Camp Gruber have not been tested or proved. Their existence, production and quality must be inferred from knowledge of adjacent areas, but would not probably produce more than 20 to 200 gpm. This may be sufficient for domestic or stock supply but would be inadequate for industrial, commercial or power-generating purposes.

River Systems

The streamflow of the Arkansas River in the vicinity of Camp Gruber is mainly regulated by the upstream Keystone Reservoir and, to a lesser degree, by the downstream Webbers Falls Dam. The average discharge of the Arkansas River near Muskogee is approximately 20,000 cfs but is highly variable and reaches a minimum flow of only 520 cfs about 1.2% of the time. The water temperatures in the river near Muskogee range from near freezing to about 86°F. The water quality of the Arkansas is poor due to a high concentration of dissolved solids of around 1,000 mg/l. This mineral content is due primarily to dissolved salt and gypsum from rocks upstream. The characteristics of other rivers in the region are summarized in Table XXXVI.

Table XXXVI. Summary of Characteristics of Other Rivers in the Camp Gruber Area

River Name	Average Flow (cfs)	Minimum Flow (cfs)	Water Quality
Grand (Neosha)	6,000	15-20	200 to 300 mg/l (good)
Verdigris	2,000	50-70	poor
Illinois	1,100	2	good
Canadian	5,600	—	moderate
Poteau	1,100	—	good

In the above table, good water quality indicates waters that meet U.S. Public Health Service Drinking Water Standards and are chemically suited for most uses. Moderate quality would be satisfactory for most industrial and energy center uses, while poor quality would usually not be used without extensive treatment.

Aquatic Ecology

The identification of potential damage to aquatic life near an energy center is an important consideration when evaluating a possible site. An analysis of the vulnerability of aquatic resources has been performed for the Camp Gruber Military Reservation. A study of the principal lakes in the area—Webbers Falls, Greenleaf Lake and Tenkiller Ferry Lake—related each water body to the local environment and the important ecosystems.

Webbers Falls Lock and Dam No. 16 is located on the Arkansas River near Camp Gruber and impounds approximately 10,900 surface acres of water. This lake lies in portions of Muskogee, Wagoner and Cherokee counties. The Arkansas, Grand and Verdigris Rivers drain into the lake along with several small streams. The lake which borders Camp Gruber was first filled in 1970. Webbers Falls lake has high turbidity due to watershed erosion and periodic dredging operations. Readings as high as 2,700 parts per million (ppm) of turbidity were observed. Visibility, as measured by a Secchi disc, is as low as 1 in. in turbid waters. The pH value of the lake ranges from 7.4 to 8.4 according to depth of the water. The average dissolved oxygen is 6.1 ppm. Soluble electrolytes, such as bicarbonates, carbonates and hydroxides, yield an estimate of productivity potential. Their measurements in the lake were quite low, averaging around 9 ppm. Dissolved solids varied between 700 and 2,140 ppm. The heavy silt loading of the Arkansas River has produced a rather low plankton productivity in the Webbers Falls Lake. Measurements of around 1,400 organisms per liter were made in 1950. The level of aquatic vegetation in the lake was also low and was attributed to the high concentration of dissolved solids, the water velocity and the steep contours of the shorelines.

In spite of the highly variable turbidity conditions of the lake from erosion runoff and periodic dredging operations, the lake is moderately clear and is capable of producing large fish populations. With stocking, it is estimated tha thte lake has a fish-carrying capacity of 600 to 800 lb/ac. Forty-seven species of fish have been identified in Webbers Lake and no species is considered rare or in danger of extinction by the Federal Register. Largemouth bass ranked first among predatory game species and bluegill ranked first in the nonpredatory game species. Carp, smallmouth buffalo and flathead catfish rank high among the food fish. All species are managed by the Oklahoma Department of Wildlife Conservation; the current practice is to emphasize the species most popular with the sport fisherman and best adapted to the lake. A vigorous stocking program was implemented in 1970 and channel catfish, largemouth bass, and walleye were introduced into the lake.

Greenleaf Lake lies about 6 miles above the Webbers Falls Lock and Dam and contains about 920 surface acres. The transparency of the water, measured by the Secchi disc, averages 3.5 ft. Dissolved oxygen ranges from 5.1 to 8.4 ppm during the summer. Plankton count varies from around 1,000 to 4,000 organisms per liter of water at the surface, to around 100 to 800 at a depth of 30 ft. The Greenleaf Lake fish resources are managed by the Oklahoma Department of Wildlife Conservation and include the same species as found in Webbers Lake. Fish sampling showed that white crappie was the dominant species with increased catches of spotted sucker and sunfishes. In 1972, estimates of the fish crop were calculated to be about 380 lb/ac.

Tenkiller Ferry Lake, located on the Illinois River, has a size of 12,500 surface acres and the water quality is rated as "good." This lake contains low-to-moderate concentrations of dissolved solids averaging 150 ppm. Nitrate levels (22 ppm) and total phosphate levels (0.35 ppm) are high, but no serious aquatic weed problems are known to exist. Nearly 40 species of fish have been reported in Tenkiller Ferry Lake, with largemouth bass adapting well, as well as both white and black varieties of crappies. Redear and bluegill sunfish are abundant throughout the reservoir waters and both reproduce well. Production of all fish species has been estimated to be 384 lb/ac.

The establishment of an energy center at Camp Gruber will have an impact on the aquatic ecology of the surrounding waters and lakes, primarily due to the large need for cooling water. Estimates of this impact will be presented in Chapter XI, on the assessment of environmental impacts.

Meteorology

Knowledge of the normal and the extremes of weather and climate factors around a potential energy center site is important for determining feasibility. Climate factors have an effect on building design criteria, cooling tower performance, air pollution dispersions and undersirable environmental effects.

The climate around Camp Gruber includes mild winters with infrequent occurrences of below-freezing temperatures and summers where temperatures often exceed 100°F. Precipitation is distributed evenly throughout the year with an average of 40 in. Spring has the greatest daily precipitation totals for Muskogee and is when flooding would most likely occur. The annual prevailing wind direction in the region is from the south with a mean wind speed of 10.6 mph. The fastest wind speed recorded nearby was 75 mph from the southwest in May 1949. Only small amounts of snow and sleet fall in the Muskogee area (7 in./yr).

Unstable or neutral atmospheric stability occurs more than 60% of the time and indicates that atmospheric dispersion is usually good.

Oklahoma is noted for its great frequency of tornadoes, which can have severe damage potential for energy facilities. Most of the damage done by a tornado's passage results from the instantaneous pressure drop, but the intense localized winds associated with the funnel cloud also have destructive power. Based on past statistics, a tornado in the site region would have an expected width of 250 yd with an expected path length of nearly 5 mi. The direction of travel would be from the southwest to northeast with a speed of 30-35 mph. The associated atmospheric-pressure drop would be on the order of 25 millibars (52.2 lb/ft^2) with tangential wind speeds of from 200 to 240 mph. The most common time of occurrence in this region is around 6:00 pm local time. The 1° latitude and longitude square enclosing the Camp Gruber site has a mean frequency of 3.2 tornadoes per year. The probability of a tornado hitting any given point in this square is 2.32 x 10^{-3}, with a recurrence interval of 430 years.

About 50 thunderstorms occur in the Muskogee area each year. The severe weather associated with these storms includes hail and lightning. Hail the size of baseballs has been reported in areas within 100 miles of the site. Lightning strikes accompany these storms and adequate precautions need to be taken to avoid storm-induced outages.

Population

Both the current and projected population distributions are concentrated westward in Muskogee and northwestward in the Tulsa metropolitan areas. Except for these two areas the region is thinly populated. The total surrounding population will rise from 644,000 in 1970 to 831,000 in 1985 and 1,016,000 in the year 2000, but will continue to be sparsely populated for the next 30 years. Forecasts indicate that the city populations will increase the most; rural areas will remain the same or increase only slightly between 1970 and 2000. Rural populations closest to Camp Gruber are expected to decline. The population impact of siting an energy center at Camp Gruber is expected to spread over a 50-mile radius rather than be concentrated. The center's peak impact on population is likely to occur during the peak of the construction activity.

Employment and Labor Force

As with any energy center, peak employment occurs during the construction phase and requires a number of skilled to highly skilled laborers. Although the region around Camp Gruber is sparsely populated, there appears to be a sufficient work force within the 50-mile radius. Camp Gruber is in the center of seven counties: Adair, Cherokee, McIntosh, Muskogee, Okmulgee, Sequoyah and Wagoner. The labor force in this seven-county area was estimated to be about 58,000 in 1970. Nonagricultural workers numbered more than 36,000, agricultural workers 10,000 with the rest self-employed and domestics. The unemployment rate reported by the Muskogee Office of the Oklahoma Employment Security Commission in 1975 was 5.8%. Greater Muskogee had an average labor force of about 23,000. The employment in Muskogee has risen consistently with the average annual rate of increase being a modest 1.7%.

The forecast employment levels for the study area are 63,000 by 1985 and up to 78,000 by the year 2000, not including any additional increase from the construction and operation of an energy center.

"Eastern Oklahoma has a history of large construction projects. Numerous hydroelectric facilities have been built on the Arkansas River system, and extensive petroleum and natural gas facilities (wells, pipelines and refineries) have been built throughout the State. It would not be difficult to find workers to construct and operate energy facilities at Camp Gruber. Recent estimates by local union officials indicate that about 9,000 construction workers are available within commuting distance of the Camp.

"Construction costs in Oklahoma are well below the national average. The recent construction cost index for the Tulsa region has been only 88% of the national average. Similarly, the recent average hourly earnings of manufacturing workers in Oklahoma has been only 92% of the national average. Construction costs are lower partly because the total amount of time lost from work in the Tulsa region has been only 32% of the national average."[2]

Recreation

Eastern Oklahoma has a variety of recreational facilities ranging from the metropolitan facilities of Tulsa to the numerous outdoor activities of the lakes and hills. Tulsa has numerous museums, art galleries, parks, gardens, zoos, theatres and other attractions. In the surrounding area, over 200,000 acres of lakes and reservoirs provide excellent boating, fishing and water sports. Hunting is very popular on the many game reserves. Extensive recreational use is now being made of the Camp Gruber area by residents of nearby communities. A survey identified 12,000 annual visits for 60,000 visitor days of recreational activity.

Historic and Natural Landmarks

There are no listed archaeological or historical sites on the Camp Gruber reservation. The closest historical sites include the original Cherokee Agency West, one mile north of the reservation, and Fort Gibson, located four miles northwest of the reservation. The Battelle study team estimated that neither of these would be significantly affected by use of Camp Gruber as an energy center.

No archaeological survey has been made of the area; however, archaeological materials dating back to 7000 B.C. have been found in the vicinity. For this reason, the study team recommends that a complete archaeological survey of the area should be made during preparation of an environmental statement written for energy facilities at Camp Gruber.

Regional Economics

An important aspect of siting energy centers is a favorable economic and industrial climate in the region. Such aspects include a receptive attitude by the local population, attractive regional economic conditions, availability of necessary support industries, and assistance from governmental and industrial development organizations. Of these factors, it is possible that the most important is a receptive attitude by the local population. Although existing attitudes may be positively influenced by a clear enunciation of the energy center's advantages and disadvantages, the current feelings of the local populace will continue to have an influence. It is encouraging that the team studying Camp Gruber had the foresight to perform a survey on the attitude of local people toward construction of an energy center. This was one of the few studies to make such a survey.

The Battelle team surveyed 900 persons residing around Camp Gruber to determine their general attitude toward construction of an energy center and the problems that they anticipate. This sample consisted of the responses from individuals living in 15 towns located from 5 to 45 mi from the proposed site. The populations of the towns included in the sample ranged from 325 to 37,331, with 38% of the respondents male and 62% female. The occupations of the respondents varied greatly with 48% employed, 2% unemployed, 21% retired and 29% housewives. Of the total 80,428 people living in this area, the sample of 900 allows an estimate of the general attitudes of the entire population to within plus or minus 5% at the three sigma level of confidence. This was a telephone poll conducted during the first two weeks in February 1975. Respondents were picked at random from telephone books and were asked a series of questions to which they could reply either yes, no or no opinion. The purpose of the poll was to determine:

1. How many people use Camp Gruber, what do they use it for and what alternatives exist?
2. How do the people in the area feel about the possible trade-offs which might be required by the construction of an energy center in Camp Gruber?
3. How do the people in the area feel about the problems that might arise from construction and operation of an energy center in the area?
4. What is their overall feeling about locating an energy center in Camp Gruber?
5. What is the respondent's occupation?

The results of the survey indicated that more residents favor (71%) than oppose energy center construction; residents of cities who have previously experienced growth were more reluctant to support an energy center; residents of communities nearer the center tended to react more favorably than residents of more distant communities. Over 50% of the area's residents would be expected to favor the energy center, even if it increased population and expanded zoning outside existing cities. Most (78%) would expect some sharing of the tax revenues which might be available from the taxing districts in which the actual construction would take place. A majority (53%) were concerned that the local communities would not be able to finance the new services required by the growth resulting from the new project. Only one-third of the respondents saw cause for concern in planning for the anticipated growth. There was a major concern expressed over possible environmental change. Most of the respondents felt there would be no problem in maintaining the area's attractiveness to tourists, but 45% said they would not favor the center if it changed the area's environmental nature. Less than half would continue to favor an energy center if it might change the environmental quality of the area. The survey showed that local residents made extensive use of Camp Gruber for recreation but that most would be willing to substitute other nearby areas. It was concluded on the basis of this preliminary survey that the local residents would be likely to accept the center. However, even if a majority of the residents have a favorable opinion toward an energy center, it is still possible that a minority of local opinion could cause costly delays in permit granting and construction which could seriously affect the feasibility of energy center siting.

Other important aspects of the region's economy are construction and operating costs, taxes and industrial development and the presence of support industries. The cost of construction in Oklahoma is well below the national average. In 1975 the construction cost index for Tulsa was about 88%. The average hourly earnings in Oklahoma have averaged only 92% of the national averages since 1960. The percent of total working time lost because of work stoppages in Oklahoma has been only 32% in recent years.

The primary taxes in Oklahoma are the usual property, sales, income and business taxes. Property taxes are used for the local school districts and other special authorizations of the taxing body. The state sales tax is 2% and communities have been authorized to impose up to an additional 2%. Business revenue or income taxes are imposed on corporations based on the amount of profit generated. It is felt that special arrangements would have to be made for certain types of business which would include the operation of an energy center.

The capital investment requirements and routine operating costs for an energy center would be smaller if the needed supporting industries were available nearby. The sponsors of the center would not have to make any investment to assure that the support industries are available, and administrative and capital costs would be distributed among more customers. In Muskogee, there are already several firms providing structural steel, metal alloys and compounds, air conditioning and heating equipment, corrugated boxes, optical equipment, and glass containers. In nearby Tulsa, there are a number of support industries including architect engineers, computer operations, consulting engineers, testing laboratories, metal fabricators, construction companies, transportation specialists and research organizations. Many of these support organizations have grown up in the area as a response to the needs of the oil and natural gas industries.[2]

SUMMARY OF CHARACTERISTICS OF CAMP GRUBER AND MICHIGAN ENERGY CENTERS

A review of the characteristics of both the Camp Gruber and the Harbor Beach area in Michigan reveals no major impediment to the construction and operation of large-scale energy centers, but it does exhibit several special characteristics of the sites which will require different solutions. At Camp Gruber it appears that sufficient cooling and makeup water will be more difficult to obtain and will require either reducing the size of the center or reallocation of the water resources from other uses. On the other hand, Harbor Beach with its plentiful Great Lakes water supply will not have this concern. Rather, there may be the serious concern for the reallocation of the scarce resource—land. By contrast, Camp Gruber contains much "surplus" land. This comparison serves to show how the characteristics of the site and the requirements of the center must be carefully matched at any proposed location.

PART TWO
IMPACTS: THE ENVIRONMENTAL, SOCIAL AND ECONOMIC EFFECTS OF ENERGY CENTERS

Part Two describes the economic, social and environmental impacts of energy centers insofar as they can be determined. Positive and negative impacts and an estimate of their magnitude will be presented for both the operation and construction phases of a center. Most of this section will deal with the impacts of the Michigan energy center concept with comparisons of other centers' impacts to depict significant differences.

IX

SOCIAL AND ECONOMIC IMPACTS
OF ENERGY CENTERS

The influx of between 100,000 and 200,000 people to an area in a relatively short time will have a profound effect on the existing social structure of the area. The impacts will be severe even with extremely careful planning and control. The existing rural social and economic structure will be entirely removed and replaced by an industrial urbanized order. Some of the present population may be willing to make the change; others will desire to find homes elsewhere; some may bitterly resist the change. Any alternatives will require accommodation, adjustment or change of the current residents which will, no doubt, be difficult. Even though the Harbor Beach area has a relatively low population density, there will still be a number of families profoundly affected. This makes the Michigan energy center different from the other proposed centers; Camp Gruber, for example, is proposed for mostly surplus lands, as are some of the other energy centers, particularly in the West.

The economic impact of the energy center at Harbor Beach will also be profound but somewhat brighter, since it is very likely to improve wage rates and reduce unemployment levels for those residents wishing to make the adjustment. The magnitude of the energy center will also have a major effect on the state, as well as local, economic picture. The total development is expected to require capital investments in the energy center, industrial center and biocomplex of around $11.5 billion in 1975 dollars. Additional capital investment will be needed for building up the urban areas of the new town. The immense size of such a project will affect the economy over a several-state region and will raise some substantial financing problems.

185

IMPACT DURING THE CONSTRUCTION PHASE

Substantial social, economic and environmental impacts can occur during the construction of an energy center which differ in character and magnitude from those of the operations phase. Environmental impacts are primarily those resulting from the construction activities and include destruction of vegetation, soil erosion and changes in the shoreline and other hydrological features. The largest social and economic impacts occur during the construction period because the labor influx will be at a maximum and the capital outlay will have its highest rate of cash flow before the center's revenues start to come in. The Michigan study team cautioned that careful planning and execution would be required to keep these undesirable construction impacts to a minimum.

IMPACT OF LABOR INFLUX

The construction of a large project involves the rotation of workers of various skills. At an energy center, the labor requirements are expected to reach a peak when 60% of the construction time has elapsed. The labor requirements then taper off. Normally this means that a temporary labor force either commutes to the site or utilizes temporary housing; both alternatives can cause special problems.

With careful planning of the construction phase, it may be possible to smooth out some of the fluctuating labor requirements. One advantage of many energy center concepts results from the relatively slow growth of up to 20 years. This will attract a more or less permanent core of workers. A second advantage arises from the construction of very similar generating units. As each specialized group of workers finishes a generating unit, they will likely be able to move on to the next unit. Some construction workers may also find jobs in the construction of the collocated industries and the center city. A smaller number of workers may be able to transfer from construction to operating jobs. The need for some construction skills will disappear slowly while the need for operating and managing personnel will increase. This means that some housing accommodations will change hands with minimum disruption. As long as careful planning and much foresight are exercised, a relatively stable work force and concomitant stable community may be maintained, another advantage of the energy center concept.

Estimation of Energy Center Labor Force

The composition and size of the Michigan energy center and new town can be used to estimate the construction schedule. Such a schedule can

can be used to estimate the likely social and economic impacts. Such estimates are also very useful for planning groups studying the feasibility of locating energy centers in other areas.

Since it is expected that the most severe social and economic impact will result from the buildup and decline of a large construction force, the Michigan study team made efforts to select a construction schedule that stretched out in time and which would limit the peak construction force required. Wide variations in the size of the force would be avoided by staggering construction of individual power plant units by one year and adjusting industrial plant construction so that it would mesh over a 20-yr period. Since some manufacturing facilities and generating units have a useful life of 15 to 25 years it may be possible to maintain a stable construction force indefinitely. Thus the earliest facilities may be near retirement age when the last planned center units are near completion. Almost continuous construction schedules, rotation of skills, retraining and shifting of construction workers to new construction or to operating positions, can all be used to assist in forming a stable continuing labor force.

Figure 30 shows the estimated labor requirements for the Michigan energy center throughout their stated construction period from 1980 to 2000. The construction labor force for the energy–generating plant is expected to build up to a peak of about 5,200 persons over a period of 8 years, to remain at this level until the initiation of the last unit, and then to be reduced to a lower level for operating and minor construction. Construction of the industrial section would occur mainly in the early years and would peak at 3,000 workers in about four years and then drop off to a steady value of about 1,000 persons. With operating personnel expected to reach a steady level of around 8,300, the combined requirements of the center would build up over a period of 16 years to a peak estimated at 13,000 workers and then drop off to a steady level of 10,300 by the year 2000.

Using Figure 30 as a basis, the Michigan study team was able to provide a rough prediction of the total influx of people into the center region. Considering only the workers directly associated with the center construction and operation, and their families, a total of 60,000 are involved. In addition there will be a construction and operating work force and their families associated with the residential, commerical and governmental sectors of the new town. A permanent population of over 100,000 will be expected around the center area; the temporary population during the construction peak will be even higher.

Current Availability of Facilities and Services

At present the population of Huron County is around 35,000. The institutional, educational and health facilities of the region are, at best,

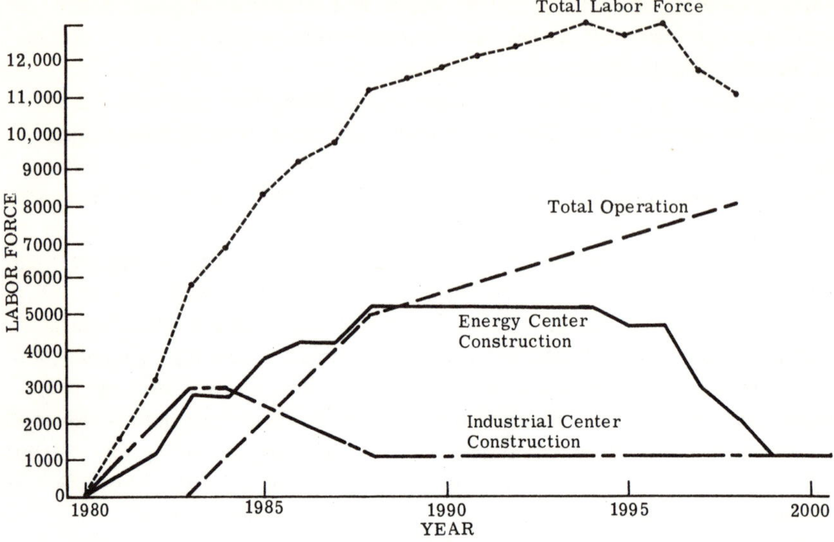

Figure 30. Construction and operating labor requirements.

adequate to handle the present population; they cannot handle an influx
of 50,000 or more additional people. Governmental services, such as police
and fire protection, are only adequate for the present rural lifestyles.
Additional services, including health, water and sewage, retail trade and
recreation, would have to be provided. These additional indirect services
combined with the population directly serving the energy center at Michigan
would be expected to create a medium-sized city of from 100,000 to
200,000 people. A comparison of the present facilities and the center–
required facilities is presented in Table XXXVII.

Table XXXVII. Present and Required Facilities and Services at Harbor Beach

Facility Capacity	Present Status	Required for Center
Educational Facilities	8,000 students	25,000
Hospital Beds	146	1,200
Physicians	26	100
Water (mgd)	2	30
Sewage (mgd)	0.35	15
Housing (Number of Units)	750 est.	33,750 est.

The major impact during the construction phase will result from the high increase in population and will severely tax the capability of existing services and the ability of the area to provide additional facilities. There will also be a greatly increased pressure for competition among different types of land use. As populations increase and additional options for land use arise, there will be increasing concern to determine the best use. This is true in Huron County in spite of the low population density. The energy center and accompanying city are estimated to require a total of about 19,000 acres; this use will have a major effect on land use in the region. At the present time a considerable portion of land around Harbor Beach is devoted to agricultural use. By the passage of the Michigan Farmlands and Open Space Preservation Act of 1974, the state is encouraging farmers to keep their land in agricultural use. Only about 1,000 acres of the energy center total is allocated for agricultural or aquacultural use, and this change in use will impose many social and economic pressures.

POSSIBLE UNDESIRABLE SOCIAL IMPACTS AT A MICHIGAN CENTER

"In the vicinity of the Harbor Beach site the current means of livelihood is predominantly farming. The sparse settlement patterns and small urban centers are typical of midwest agriculture. The influx of permanent construction workers, engineers and operating personnel will demand not only an increase in the quantity of public and commerical services, but also an increase in quality of such institutions and facilities as schools, hospitals and the transportation network. Traditionally, the cost of such improvements and additions has been borne by the local government (with some state and federal grant assistance). Experience shows, however, that even the best of planning and anticipation by the local authority may be thwarted by the pressures of development generated in the public market place, which are not under direct control. Much can be done to alleviate this through more sophisticated planning and use of the powers local governments have. New and innovative designs for planned development must be devised to avoid degradation of the natural and cultural environment by creation of undesirable urban sprawl, commercial strips, concomitant inefficiencies in use of the land and difficult-to-control suburban inflation, where the local government is always two steps behind the public demand for services.

"Unless planned otherwise, the residential development to house the expected influx of workers will most likely occur first along the lake front in a tightly packed strip which could block access to this resource to others. The wages of the various types of workers will allow ownership of modest single family homes, and the pressure will be to provide these. As soon as the first contingent of construction and operations people arrive and the settlement pattern is established, the associated service businesses will establish themselves adjacent to the residential areas—the fast food shops, movie houses and small retailers. After a few years, there will be a pressure to construct one or more large shopping centers, as the total population of the area increases. The rise in the number of people not directly associated with the energy center will lead to construction of modestly priced apartments around these centers.

"Along with these changes will come the complete disruption of the existing small urban centers. Harbor Beach, the closest, and by the lake, will be hit by the first construction workers. Temporary motel housing may be required, and a plethora of such establishments will be built. As permanent housing becomes available, the market for these motels will disappear, forcing them out of business. As the outlying commerical development expands, the small towns will be replaced by the shopping centers and commerical strips.

"Since the county, and not an established town, may have jurisdiction over much of the area to be developed, it will be expected to bear the brunt of providing the public services—schools, water and sewage lines, hospitals. However, residential and commerical development can occur at a rate an order of magnitude faster than the financial and even physical capability of the local government grows to meet these needs. Initially, the existing financial tax base of the area will not support the necessary water and sewage systems required, and the first wave of residential and commercial buildings will rely upon wells and septic systems. Drainage in the area of the Harbor Beach site is poor, and the land's ability to sustain septic systems is probably very limited. This could begin a rapid degradation of the quality of the available water table and surface waters both and a lessening of the quantity. The tax base will not allow construction of the required schools and other public services; thus, the existing facilities will be overcrowded and the quality will deteriorate. These situations eventually will be alleviated as the population stabilizes but the harm done to both the natural and social environments could be irreversible.

"The above sequence has been followed many times in urban and suburban environments. A Michigan energy center would offer a unique opportunity to exercise planning techniques to try to avoid these problems, which so far have plagued all of our urban areas. Avoiding these problems will require, however, not only good advanced planning, but some perhaps significant changes in the current power structure between state and local governments and their ability to control private development. Currently, the only direct control available lies with the local government, through zoning. It is difficult or impossible for a local authority to anticipate and prepare for such a major development as the energy center, because of insufficiencies in both funding and staff, and lack of technical expertise. Some type of joint effort between state and local agencies, probably with assistance from federal funds, will be required to properly anticipate and plan for the major development brought by an energy center of this magnitude.

"Even given good multilevel cooperation, the ability of any level of government to control private development is limited. Existing zoning ordinances, which are the only direct method available, are in general weak and fairly readily circumvented. Indirect methods have been proposed, but these usually rely on establishing public services, such as sewers and schools, before development occurs. They also assume that the private developer will follow these in his choice of location. Obviously, though, this requires major capital investment before the tax base exists to provide it.

"The whole idea of planned development to optimize and preserve the quality of the social and natural environment associated with an energy center is one that obviously needs to be studied in much greater depth. Many of the problems and questions are political and legal ones that need to be anticipated well in advance, and potential solutions to them need to be found before they become moot. It is not known to what degree the public of Michigan will accept a planned community, and trade-offs between the perceived loss of individual choice and benefits gained in improved quality of life and efficient use of resources; hence, this subject must be explored."

IMPACT ON POPULATION AND
REGIONAL ECONOMICS

A considerable emigration and population decrease in Huron County is presently occurring because of the increased mechanization of farms and because of the higher wages obtained elsewhere. This trend is likely to continue and perhaps increase if industrialization comes to the area as the result of an energy center. This will no doubt result in an almost complete turnover of the population since the energy center will require highly specialized and skilled construction workers which are presently not available in the region. The impact of such a complete turnover of population cannot be estimated, but it would be erroneous to assume that because the overall or average economics of the county would be improved by center construction that the current economic problems of the area would be solved. The problems would not be solved by relocating them. Yet, the dislocation might be alleviated to some extent by the increased opportunities in local commerce and services needed to support the new community. The biocomplex to be located at the Michigan center may provide a job market for some of the present inhabitants of Huron County.

SOCIAL AND ECONOMIC IMPACT OF THE AGRICULTURAL APPLICATIONS

The location of an energy center at or near Harbor Beach would replace some prime-to-fair farmland with a concomitant reduction of farm population and production. The impact will be reduced if some of the waste heat can be dissipated through agricultural uses. In the Michigan study, it is proposed to have 960 ac of labor-intensive high-productivity agriculture.

The methods by which rights to the necessary land base are acquired and the current local economy will have the most effect on the social and economic status of the adjacent area.

Some of the options for acquiring the rights to the land for an energy center are given in Table XXXVIII.

Table XXXVIII. Site Acquisition Options

Fee Simple Acquisition
    Purchase and Manage
    Purchase and Leaseback
    Purchase and Resale on Condition

Less Than Fee Simple Acquisition
    Purchase Easements

Contractual Agreements—For Rent Property Interest
    Waste Heat Water Cooperative
    Contractual Arrangement

Public Authority

These various options differ in the capital requirements, degree of control exercised over the system, managerial skills required, impacts on the local community, engineering design, and income generation which the utility companies must consider. For example, the costs incurred by a company seeking fee simple acquisition would include payment of interest and principal on bonds raised to finance purchase of the land, administrative costs and cost of compensating the affected communities for property taxes foregone where land is purchased by a tax-exempt body and leased back for agriculture and/or aquaculture purposes. On the other hand, should the utility company decide not to fully control the total integrated system and not need to raise the capital for some or all of the separate subsystems, it could enter into a contractual agreement to supply waste heat water to private concerns. At the extreme, the utility system could be publicly owned and managed, like the Tennessee Valley Authority. Many diverse criteria would be used to evaluate these options: considerations would include the effects of soil warming and laying of pipe on the land, flexibility and permanency of access to the land, goals of the community, political sensitivity to land acquisition, cost of the purchased land and how the degree of control over land use might affect the goals of utility.

ECONOMIC EFFECTS

The agricultural and aquacultural systems would directly provide only about 300 full-time jobs when in full operation. However, this is not the only measure of the overall effects of such a project. There is also the impact of construction and the added stimulus on the local economic community during the period of operation. Among the additional effects will be jobs created for marketing the products, aquiring capital, manufacturing system components such as pipes and pumps, warehousing and many others. Another positive economic impact of the agri/aquaculture systems accrues from the substitution of some of the natural draft cooling towers at the site which would result in this use of waste heat. Such savings have been estimated at having a present worth of about $53 million based on the assumption of a 12% discount rate for future revenues and costs and a 30-yr lifetime.

A major problem to be faced in the development of a large energy center complex is the need for large amounts of long-term financing. The basic problem of obtaining adequate financing for future power needs is common to all power developments but takes on special importance for an energy center because of the concentration in time and location compared to dispersed siting. It appears that the vast size of the energy centers will require special treatment if development is to occur. It is likely that no single private industry could generate the needed capital and that state or federal financing, or some combination of the two, will be required.

The solution to the financing problem will be directly related to the type of organization formed to manage the enterprise. For example, a public authority corporation would obtain major amounts of its financing from the federal government. Private corporations may get assistance from government units, but would receive much of its financing from the private sector money markets. This distinction between public and private financing is not only a matter of economics, it also has important political and social implications. Most of the energy center studies, including the Michigan study, have not performed a detailed analysis of this most important topic.

FINANCING THE COMPLETE ENERGY CENTER

The energy center concept consists of a variety of components—utility, industrial, agricultural, urban and institutional. Although different sources of financing might well be used for each of these components, the success of the total system will depend upon the availability and functioning of all of its separate components. A smooth symbiotic relation among the individual components of the system implies that there must be assurance of adequate financing for the entire system at the beginning.

Most of the capital needs are required for construction of the power generation and distribution system. In recent years, the condition of the usual financial markets has caused increased difficulties for the electric utilities to obtain adequate financing for continued expansion. Early in the construction phase of the Michigan Center, yearly capital requirements may be as high as $1.4 billion. By way of comparison, yearly capital expenditures by the Michigan utilities were running about $500 million in 1975. Raising this amount of capital for new facilities has been difficult. Tripling the amount of capital to be raised by private industry might be even more difficult in the future. Those industrial companies planning to cooperate in the development of the energy center complex will face similar financing problems. This problem may be less severe if the prospects for economic and energy conservation advantages to be derived from the energy center development materialize as planned.

Arguments in favor of public financing derive from the ability of the federal and state governments to borrow funds at lower cost than private organizations; this would reduce the total system cost. Government participation in the financing could come in a variety of forms, including tax incentives for private industry, subsidies for essential public services and furnishing or guaranteeing funds needed for the center development. The advantages of public financing are questioned by those favoring private financing, who claim that some costs are hidden and shifted to taxpayers. There is also concern over the political implications of the shift of more control over the economic system to federal or state governments.

Most of the energy center studies conducted to date are preliminary and do not offer solutions to the financing problems. Nevertheless, they all point up the severity of the problem which results from the large scope of the proposed systems.

X

PUBLIC ATTITUDES

Several large energy facilities have recently suffered many additional and unexpected delays in construction that have increased costs way above expectations. Some of these delays have been caused by the objections and criticisms of local residents. For energy centers, their large size and the already crucial problem of obtaining sufficient financing may produce an adverse leverage effect when small groups of adversaries can produce costly delays. A major positive advantage at energy centers is their overall efficiency and economic improvements. However, these advantages in economies-of-scale are marginal; an unexpected increase in construction costs from delays may well destroy most of the advantage. It is imperative, then, that a region exhibit a strong favorable public for the siting of an energy center. This is only a minimum requirement on public attitudes. There is the possibility that small special interest groups, while not representing the majority, may still create sufficient objection to cause expensive delays. The highly probable fact that different people in a given region will favor different values demands that these attitudes be investigated and their effects predicted and factored into the energy center feasibility studies.

The Camp Gruber study team recognized the importance of assessing local public opinion. They performed a public attitude poll as part of their initial feasibility study. The results of this poll have already been summarized in Chapter VIII.

Michigan has, of course, a number of constituencies with differing economic, social and environmental values and goals. In recent years, various specific economic and environmental issues have been debated among private groups, public officials and interested citizens. This discussion has resulted in much environmental legislation being passed with the objectives of protecting air and water quality, controlling land use and protecting public health. It is difficult to compare strengths of the opposing

195

opinions and the outcomes of debates over energy issues. The Michigan study group performed some tentative evaluation of the composition, attitudes and numerical strengths of individual constituencies and made a brief survey of public and legislative opinion. A summary of their findings is presented in the following section.

INTEREST GROUPS IN MICHIGAN

The environmentally oriented segments of Michigan citizens consist of many diverse elements organized at both the state and federal level. Environmental interests are promoted not only by these groups but also by legislative and executive personnel of the state government, professionals in law and natural resources, editorial staffs of some newspapers, certain labor unions, students and many others. Other interest groups, including manufacturing organizations and labor unions, are concerned about the state's economic health.

Michigan industry is diversified, but a major portion is concentrated on the manufacture of motor vehicles. Historically, Michigan's economy has been susceptible to the cyclical nature of the automotive industry. The construction of one or more energy centers would help diversify and stabilize the state's economy, and would probably receive favorable attention from certain industrial organizations and commerce groups in the state government. Many Michigan labor unions are favorably disposed toward increased utility plant construction and industrial expansion, although not necessarily for an energy center as such. This attitude is illustrated by the April 1975 decision of the Greater Detroit Building Trades Council and the Michigan State Building Trades Council to intervene before the Michigan Public Service Commission on behalf of Detroit Edison's rate increase request. The suggested rate increase would alleviate the financial problems that caused postponement of the $6.4 billion proposed power plant construction by Detroit Edison and Consumers Power. Since the construction and operation of energy center in Michigan would create many additional job opportunities for the region, it is likely that these trade organizations would be in favor of the concept.

The people who might be affected adversely by the center's construction are the local residents near the site. This segment of the public is likely to be influenced by a variety of factors, and it will be very difficult to estimate the impact or the response of these people in advance of construction. From a purely monetary standpoint, it might be expected that property owners in the site area would do well. However, their interest may go beyond the matter of obtaining adequate compensation for their land. In the Harbor Beach site area, many of the residents are farmers. In talking with some of them, it was learned that it would be extremely difficult to reestablish a comparable farm elsewhere, and many of these people have a strong attachment to their current landholdings. This seems particularly true of the older farmers who would not find it easy to start again. It is a difficult moral question to consider—placing great hardships on a few residents for the benefit of many others. These human factors involved in allocating land for energy center construction therefore need to be weighed very carefully.

Some insight into the attitudes of the general Michigan public toward electric energy policy can be obtained from a survey conducted by National Marketing Surveys of Midland, Michigan, which interviewed over 2,000 voters randomly selected to represent each of the 38 State Senatorial Districts.[40] The following statements represent some of the conclusions reached from the survey and indicate probable public attitudes concerning energy center development in Michigan.

The voting public is genuinely concerned about the future availability of electric power; 52% are convinced that a chronic power shortage is likely within the next five years. Nearly 71% foresee a shortage that adversely influences job opportunities and levels of employment; 64% foresee the imposition of measures leading to personal inconveniences. The public is prepared to make private and public accommodation to insure adequate energy supplies. "On a community level, a majority of people express willingness to accept either a coal-burning plant (79%) or a nuclear plant (64%) if needed to assure adequate electric power. Surprisingly, persons who are supportive of environmental organizations display a slightly greater acceptance of these plants than does the general public. Within this select group (6% of the total population), 80% would welcome a coal-burning plant and 71% a nuclear plant. These plants would be even more appealing if there were prospects of reduced electric rates or lowered taxes, in which case less than 9% of residents would be totally unreceptive to both types of plant."

RECENT HISTORICAL ISSUES

Although it is difficult to generalize from the results of specific events, it is often true that the opposing interests reach some sort of compromise. The history of three recent environmental issues illustrates this tendency.

Oil and Gas Drilling

A decision on whether to allow drilling for oil and gas in the Pigeon River County State Forest in northern lower Michigan is a classic case of the conflict between environmental and economic goals in Michigan.

State geologists estimate that this area contains more than 140 million barrels of oil worth $1.6 billion and natural gas worth another $108 million. Substantial income would go to the state of Michigan. The controversy centers on the threat to wildlife in the area, particularly to Michigan's dwindling elk herd. Also at issue is the danger of general disturbance to an environmentally sensitive and scenic area. Several energy companies, including Shell Oil, Standard Oil of Indiana and Consumers Power Company had taken 10-year leases on the mineral rights in the area in 1968, but have so far not been permitted to drill for oil. Several methods of oil drilling have been outlined, some of which would limit the amount of actual drilling required.

After several delays and court challenges, Dr. Howard Tanner, Director of the Michigan Department of Natural Resources, set up a DNR task force to prepare an environmental impact statement. This statement was distributed by the DNR to the public and several public hearings were held in the state to determine opinions and reactions to the proposal. The environmental impact statement analyzed two alternatives: to permit no drilling or to allow such production only in the southern third of the area. This latter plan was the most restrictive of the three proposed drilling plans. (The third being no restriction on drilling.)

The implications derived from this situation are that substantial delays, court challenges and the concomitant costs are likely to result from the opposing objectives of the economic and ecological interests involved. It is encouraging to note that evidence so far shows that state government units are sensitive to the arguments of both sides and that a balanced decision is likely to occur.

Steel Plant Siting

Another illustration of the opposing interests of economy and ecology was the proposed plan by North Star Steel Company to select the Muskegon area as a site to build a $50 million mini-steel plant, which would employ 500 workers. One opposition group, called Save Our Shorelines, filed a federal lawsuit to block the company's plan to fill 56 acres of Muskegon Lake for the plant site. It is interesting to note that this legal action was based on federal rather than state law. In addition, delays by the U.S. Army Corps of Engineers would hold up the project for around nine months so that an environmental impact statement could be prepared. It was the Company's contention that each month's delay would cost about $1 million. Because of the opposition, it decided to drop construction plans at Muskegon and look for an alternate site either in Michigan or elsewhere. The North Star decision may have a considerable impact on political attitudes in Michigan concerning the relations of economic vs environmental advantages. The decision seems to have dissappointed the Office of Economic Expansion and several state legislators from economically depressed districts, labor leaders and some business associations. It may be impossible to evaluate the wisdom of North Star's decision for some time to come.

Modification of the Michigan Environmental Protection Act

Another issue which has been the center of a highly political controversy, and which has implications for siting energy centers in Michigan, involves the possible modification of the Michigan Environmental Protection Act of 1970. This bill would prohibit Environmental Protection Act suits from blocking mining projects once all required governmental permits for them had been obtained and 30 days had elapsed. The impetus for writing this bill was the prolonged delays in constructing a coal dock in Marguette Bay in the Upper Peninsula, a region of historical economic deprivation. The dock was designed to speed deliveries of low-cost water-borne coal to a planned electric generating plant serving a $1 billion iron mining and ore pelletizing project. Environmental groups have opposed the bill on the grounds that it would not only endanger the Upper Peninsula's environment but also seriously weaken the state's Environmental Protection Act. The modification bill initially passed the State Senate, but substantial opposition developed and it was decisively defeated in the House.

OBSERVATIONS ON PUBLIC ATTITUDES

Although it is difficult to foresee the outcome of future controversies between the environment and the economic elements based on past occurrences and the present climate of public opinion, it is possible to form a few general conclusions. It appears that the opposing sides on the energy center issue may be rather evenly divided in numbers, if not in effect. The economic interests in favor of an energy center would most likely include business, labor and the financial sectors. On the other side would be environmentalists, some landholders in the site proper, and perhaps some labor groups. There is probably no way to determine the disposition of the local residents to a site until formal plans are proposed; they will be divided between those welcoming the center construction because of the economic benefits and those with an attachment to their existing way of life.

Based on similar endeavors, it may be safely predicted that any legislative changes needed to relax environmental requirements applying to center construction will meet determined opposition. It is certain, given the current climate of opinion, that legal challenges to energy center construction will reach the courts in some form and will take a long time, and much money, to resolve. This single factor may be the most serious impediment to the feasibility of large-scale energy-generating facilities at one location, but it is not necessarily a lost cause. Public approval for an energy center will depend on how convincing its proponents are, the current economic climate of the proposed region, the care taken in its planning, and a demonstrated determination to minimize environmental damage. In summary, it will depend on the balance of economic need for the facility against cost of protecting the environment.

With the existing environmental climate in Michigan, opposing action can be predicted to emerge in three general forms: (1) to challenge the need for an energy center and the assumptions upon which the need is based; (2) to monitor implementation of state and federal environmental statutes; and (3) to insist on citizen participation in decision-making at all steps.

These would not be unreasonable expectations, in fact, they would be identical to the objectives of those involved in constructing the center. It is the author's opinion that early citizen involvement in energy center planning would not only assist in demonstrating its feasibility (if in fact it is), but would also reduce the impact of the deleterious and irrational environmental fringe, the so-called holdouts. Given the advantages and disadvantages in an unbiased manner, the people can make the right decision.

XI

ENVIRONMENTAL IMPACT OF ENERGY CENTERS

The several studies of energy centers, as opposed to the dispersed siting of an equivalent generating capacity, have revealed no additional effects of the center on the environment. That is, the total environmental impact of an energy center would be no more severe than if the same facilities were sited in a dispersed manner. The distinction between the alternatives is that for the center the impact occurs at one location. For some factors, this concentrated effect is an advantage, since the impact is confined and more easily cared for. For other factors, the results of congregating facilities can cause severe difficulties. This section discusses those environmental impact that will be worsened or eased by congregating the effects. Most of the discussion derives from the Michigan study experience. Those impacts which will occur regardless of whether the facilities are compacted or dispersed will be discussed only briefly.

WATER QUALITY

Water Quality Standards

In all energy center studies conducted to date the question of water availability and the effect on water quality have been paramount. For example, the state of Michigan has established a set of water quality requirements, applicable to the Great Lakes, their connecting waterways and all other surface waters, which have implications for siting energy centers. The purpose of these standards is to protect the public health and welfare and to protect the quality of waters for recreational purposes, public and industrial supplies, agriculture uses, navigation, and the propagation of fish and wildlife. The Michigan water quality standards were approved on September 21, 1974, as part of the Michigan Administrative Code. The Great Lakes and some of its tributaries have multiple uses and are thus protected with the most restrictive water quality standards. The Michigan law also contains the following nondegradation and water quality improvement clause[1]:

> Waters of the state in which the existing water quality is better than the water quality standards prescribed by these rules on the date when the standards become effective, shall not be lowered in quality by action of the commission unless and until it has been affirmatively demonstrated to the commission that a change in quality will not become injurious to the public health, safety or welfare; or become injurious to the domestic, commercial, industrial, agricultural, recreational or other uses which are being made of the waters; or become injurious to livestock, wild animals, birds, aquatic life or plants, or the growth or propagation thereof be prevented or injuriously affected; or whereby the value of fish or game may be destroyed or impaired; and that a lowering in quality will not be unreasonable and against the public interest in view of the existing conditions in any waters of the state.
>
> Waters of the state which do not meet the water quality standards prescribed by these rules shall be improved to meet those standards. Where the water quality of certain waters of the state do not meet the water quality standards as a result of natural causes or conditions no further reduction of water quality by controllable point and nonpoint sources shall be permitted.

The state standards have been set to control the amounts of suspended solids, dissolved solids, pH, taste- and odor-producing substances, toxic substances, radioactive substances, plant nutrients, fecal coliforms, dissolved oxygen, temperature, impoundments, mixing zones and dredging. The water quality standards impose on a Michigan energy center, using the best available technology, the control of all sources of contaminants and pollutants and the minimization of their release to the environment by all possible pathways. These necessary controls have associated costs that must be taken into account when determining the feasibility of the center.

Impact on Water Quality from an Energy Center

A Michigan energy center would have a substantial impact on the water quality of the Great Lakes and local receiving waters. The principal sources of this impact will be from (a) fallout on the lakes from the atmospheric pollutants discharged by the center; (b) soil erosion causing sedimentation of the streams and lakes, especially during the construction phase; (c) groundwater contamination from center waters; (d) surface storm runoff, and (e) the discharge of heated power plant cooling water into the Great Lakes. Each of these pathways must be considered carefully if the environmental impact of the energy center is to be minimized.

The national water quality goal is for zero discharge by the year 1985. To meet this goal, a center must use the best available technology for the treatment of wastewater from the power, industrial and domestic sources. This goal can best be achieved at an energy center by considering closed-cycle systems. In such a system, the wastewater would be treated to a quality level sufficient for reuse. When such systems are incorporated into the energy center complex, the problem is largely reduced to one of solid waste disposal. Some of these solid wastes might be recycled for both center and noncenter uses.

Center-related activities may constitute a hazard to water quality from the possibility of oil-contaminated discharges and catastrophic spills. Such spills from an energy center might include heavy metals harmful to aquatic life. In the past it has been found that these metals form organic contaminants which are resistant to natural breakdown and tend to accumulate in fish. These materials usually enter the food chain through benthic organisms. Mercury is one such example which formed an organic compound and was detected in both Lake St. Clair and Lake Erie in 1970.

Potentially adverse impacts can occur from the runoffs from agricultural activities of the center. This causes a release of nutrients which causes a condition of overenrichment (eutrophication) which stimulates aquatic plant growth. Other adverse effects include an increased turbidity from construction activities, increased biological and chemical oxygen demand, and changes in the pH level. High concentrations of dissolved oxygen are necessary to sustain a favorable aquatic ecosystem.

The center, with its highly concentrated power facilities and industry, would have a much greater impact on local water quality than dispersed siting of the same capacity. It is doubtful if the total impact would be greater, but by being concentrated it would certainly be more noticeable. This does not make dispersed siting more desirable just because it might go unnoticed. In fact, at an energy center, it will be necessary to take more extensive precautions in order to provide an acceptable impact on local water quality and, therefore, possibly have less total adverse impact than dispersed siting.

Treatment of Paper Mill Wastes

Present treatment systems for paper mills provide for the removal of suspended solids and biologically oxidizable materials using settling basins, activated sludge processes or trickling filters. The effluent quality obtained by these current methods will not be acceptable for the energy center mill feedwater requirements. The main problem is that these systems do not remove the color from the effluent. They may have to be augmented by activated carbon absorption and lime precipitation.

Effluent flows for a paper mill range up to 60,000 gal/ton. This rate might be reduced to 4,000-6,000 gal/ton through improved design and increased in-plant reuse without the necessity of the removing dissolved organics. A current effort is being made to use precipitation and biological oxidation for mill treatment and recycle. Many of the advanced treatment processes being investigated include the traditional biological oxidation process. Some of the suggested treatment processes for water recycle would require large land areas and would entail increased land acquisition costs. For example, nonaerated stabilization ponds would require 40 ac for each mgd capacity. With aeration, the pond area requirements would be reduced to 2 ac/mgd.

Treatment of Refinery Wastes

As the criteria on effluent purity imposed by various regulatory agencies become more stringent, water recycling and reuse in refineries will become more attractive. One of the major problems associated with the treatment of refinery waste is that the various waters used will have widely varying levels of contamination. Additionally, for refineries sited near the Great Lakes effluent quality requirements may be more stringent than those acceptable for other areas. An EPA report[41] concludes: "Complete treatment of refinery effluents by physical, chemical and biological means is possible although treatment is usually both expensive and problematic. Nevertheless, water reuse should be incorporated into existing plants and made an integral part of proposed plants."

Stormwater Runoff

In addition to the treatment of waters used directly in the various industrial processes, any effective energy center wastewater control program must address the problems of contaminated storm runoff. Since a large area of the center will have impervious surfaces, most of the precipitation falling on the center will end up as runoff. Any exposed soil areas could allow some infiltration to, and possibly some contamination of, groundwater supplies as the precipitation washes the facility. Most large-area industries, such as refineries, include in plant designs dikes and catchments of sufficient size to collect storm runoff. Often such runoff waters can be used with little or no treatment.

For an energy center as planned, it may be desirable and practical to collect and treat as much of the storm runoff as possible. It then would become necessary to determine how contaminated the stormwater will become, how much of it will recharge the groundwater, and how much will escape the center as runoff. Any significant amounts of the latter would be unacceptable because of the adverse effects on the water quality of the nearby Great Lakes.

In order to assess the problems of runoff, one needs information on: (1) the hydrology of the center site; (2) the ratio of impervious area to exposed soil at and near the center; (3) the quality of stormwater runoff from various center activities; and (4) the characteristics and uses of groundwater in and around the site area.

The lack of quality stormwater runoff is very significant and is comparable to raw sewage effluent. Biological and chemical oxidation demand is comparable to that of typical secondary sewage treatment effluent, while suspended solids concentrations are comparable to raw sewage. Comparisons are made in Tables XXXIX and XL. Other studies of urban runoff have identified the major sources of pollutants as materials deposited on impervious surfaces and material eroded or picked up from drainage channels. Other sources include materials from catch basins, roof discharges and chemical storage piles. The land characteristics most responsible for the amount of stormwater pollutants are environmental conditions, geomorphic drainage characteristics, land development and population density.

The impact of stormwater runoff from an energy center complex on receiving waters should be similar to the cited studies which deal with street runoff in urban areas. Heavy metals such as zinc, cadmium, mercury and lead may be more predominant from an energy center, and may be concentrated in sufficient quantities to cause toxic effects to aquatic organisms. Such contamination would have a serious effect of the littoral waters of Lake Huron or Lake Michigan in the vicinity of the center sites.

Control measures to reduce the adverse effects of storm runoff include catch basins, sludge removal, sweeping of impervious areas, stringent standards for particulate release, land drainage improvements, and stormwater

Table XXXIX. Comparison of Constituents of Stormwater Runoff with Domestic Sewage[42]

| Constituents | Raw Domestic Sewage | | Urban Runoff Loads as Percentage of Sewage Loads | |
	(lb/day/ac)	(lb/yr/ac)	During Runoff	Annually
Suspended Solids	1.5	540	2,400	160
COD	2.6	960	520	33
BOD$_5$	1.5	540	110	7
Total PO$_4$	0.19	68	70	5
Total N	0.23	82	200	14

Table XL. Calculated Quantities of Pollutants Entering Receiving Waters in a Hypothetical City[43],[a]

Pollutants	Street Surface Runoff Following 1-hr Storm (lb/hr)	Raw Sanitary Sewage (lb/hr)	Secondary Plant Effluent (lb/hr)
Settleable plus			
Suspended Solids	560,000	1,300	130
BOD	5,600	1,100	110
COD	13,000	1,200	120
Kjeldahl Nitrogen	880	210	20
Phosphates	440	50	2.5
Total Coliform			
Bacteria (org/hr)	$4,000 \times 10^{10}$	$460,000 \times 10^{10}$	4.6×10^{10}

Note: Since the above calculations were based only on a five-day accumulation of street litter, the above discharge of contaminated runoff could conceivably occur many times in a year.

[a]The hypothetical city has the following characteristics:
  Population: 100,000
  Total land area: 14,000 ac
  Land use distribution:
    residential: 75%
    commercial: 5%
    industrial: 20%
  Streets (tributary to receiving waters): 400 curb mi
  Sanitary sewage: 12 mgd

treatment using the best available technology. Based on an EPA report,[42] the annual contributions from storm runoff of impervious surfaces on the industrial uses of a 5-mi^2 concentrated center could include 15 tons of zinc, 1.7 tons of nickel, 27 tons of manganese, 55 tons of lead, 135 tons of iron, 5 tons of copper, 13 tons of chromium, 59 tons of orthophosphate and 78 tons of nitrogen. It is interesting to note that the phosphorus and nitrogen losses would be less than those associated with an equivalent area engaged in agriculture. The quantities given for the above elements represent an average for industry. With a good design, the center could be much cleaner than the average industry. Still this is a very important problem and the center design will have to deal with the severe problems of stormwater runoff and resulting contamination of local waters. If the plan includes the retention and recycling of the center's stormwater, several acres of ponds will be required to catch the largest storms; this land requirements must be included in the center feasibility study.

THE IMPACT OF THERMAL DISCHARGES

The effects released waste heat from electric power generation to the Great Lakes waters is still an unsettled question. The environmental impacts of waste heat are very hard to determine because heat is not persistent and does not accumulate like chemical pollutants. On the other hand, this lack of persistence is likely to make thermal pollution a less serious problem.

The current EPA standard of thermal discharge falls under the national water quality program of 1985. It has been stated that all power plants must incorporate the best available technology to minimize thermal discharge as well as other pollutants. The "best available technology" often seems to be "the technology currently proposed." The best available technology that is presently being promoted is closed-system cooling. The alternative for energy centers on the Great Lakes is the once-through system which utilizes the almost infinite heat sink capability of these great bodies of water. The possible meterological alteration caused by direct atmospheric discharge of heat in a closed system may, in the long run, have a greater adverse effect than once-through colling. There is no doubt that a large discharge of heat to either the atmosphere or the Great Lakes has the potential for causing harm. More study is required to discover which approach will produce the least adverse effects.

Discharge to the Great Lakes

The natural thermal assimilative capacity of the Great Lakes is a major resource and should be utilized, providing detrimental environmental impact can be avoided or at least reduced. As in all technological endeavors, trade-offs between good effects and bad effects must be carefully weighed; it is impossible to do just one thing.

The primary concern over thermal pollution is its effect on aquatic organisms. In Michigan, it is required that an applicant for thermal discharge obtain a National Pollutant Discharge Elimination System Permit from the Michigan Water Resources Commission. These permits are granted in the form of a mixing-zone criterion. Such a criterion defines the size and shape of the thermal bloom that extends in the receiving waters. The size and shape of the mixing zone is defined by the applicant. The increase in temperature at the edge of the mixing zone cannot exceed natural water temperature by more than $3°F$. In Michigan, the area cannot exceed a circle of 1,000-ft radius unless it can be shown that the defined area is more stringent than necessary to assure protection of aquatic life.

Even though it can be proven that the requirement is too stringent, it is still required (as in the case of the Donald C. Cook Plant near Bridgman, Michigan) that the applicant develop a comprehensive monitoring program to assess the long-term impacts of the plant's operation on Lake Michigan. Monitoring includes measurements of thermal plume temperatures and the effects on aquatic life, including fish, zooplankton, phytoplankton and benthos. The outcome of these long-term studies under operating conditions will increase our understanding of the effects of once-through cooling using the Great Lakes as a heat sink.

Extrapolating from the mixing zone used by the Donald C. Cook plant on Lake Michigan, an energy center would require an area of 6,218 ac, equivalent to a circular area with a radius of 9,285 ft. This mixing zone is 7.5 times the size of that permitted at the Cook plant; it is uncertain at this time what adverse effects, if any, would occur.

The alternative to once-through cooling is the closed cooling system using ponds and cooling towers. In this case, the waste heat is discharged into the atmosphere. Although this method has become the preferred method of dissipating heat, it is possible that the potential long-term effects on the atmosphere would be more serious than those experienced by once-through cooling. The concentrated release of both heat and moisture into the atmosphere from an energy center may result in significant meteorological impact.

Because of the uncertain impact associated with both atmospheric release of water vapor in a closed system and the thermal effects of once-through cooling on the lakes, it is possible that some combination of the two methods would be best. Therefore, siting considerations should allow both types of system to be implemented. As the center develops, tests for the effects can be made so as to determine which approach is producing the least effect on the air or water quality. The large amounts of heat generated may be partially utilized by the associated industries and agricultural applications; however, it appears at this time that most of the heat will eventually escape to the atmosphere or to the Great Lakes.

Allocation of Biological Value

There are limits to the encroachment of thermal and waste discharges into areas of the Great Lakes. The Lakes support significant biological resources and this encroachment can cause an ecological imbalance and eventual population decline and collapse. Areas sensitive to encroachment include spawning sites, nursery areas and food-producing areas. The EPA has proposed a limit on the encroachment of thermal and waste discharges in terms of the loss of biological value to the receiving waters. This

approach may be more reasonable than an arbitrary mixing zone area. However, regulatory agencies must agree on the biological and other uses protected. It is likely that many conflicts in the allocation of resources will occur in the planning of the Great Lakes Region. Thus, the relative biological value for various biotic zones must be determined in order to allocate to them portions of each zone. Portions of the mixing zones will be allocated to discharges from paper mills and municipal wastes besides the power plants and other industries.

Regulatory agencies must make a variety of individual studies in each area. Species important to the stability of the ecosystem must be identified and their biotic zones must be mapped; maps of the thermal discharge would determine the mixing zones that will minimize loss of biological value. Plant design difficulties or land siting constraints must be balanced against the biological loss and a scale of values must be assigned to form a compromise in the various biotic zones.

A list of values have been suggested by the International Joint Commission for the Great Lakes. The values were assigned on the basis of the relative importance of each area for the maintenance of balanced ecosystems.

Open water greater than 100-ft depth	1
Exposed sand shores to 20-ft	2
Waters with depths between 20 and 100 ft	3
Nursery areas	7
Protected rock shores	5
Spawning grounds	10
Unique spawning grounds	infinite

Future mixing zones, of the size and impact which is contemplated for an energy center, should obviously be located in the less valuable biotic zones. Fragile systems with low but important productivity might require a very high level of protection. For ecosystems with a capacity for regeneration and with the ability to withstand impact, perhaps as much as 8% of the total biological value could be assigned. The discharge of the center will be allotted a total biological value which could be the simple product of the mixing zone radius times the biotic zone value.

Once-through cooling for an energy center of the size considered in Michigan will require a very large mixing zone. Therefore, the center would have to be allocated a ratio of large biological value. This requirement would be more easily met if the mixing zone could be sited in an area of low biological values. Mixing zones would probably have to be in waters with depths in excess of 100 feet. For the Michigan site at Harbor Beach, these depths are obtained at about 3 miles offshore. No doubt

the warm discharge water of the center could be pumped out into the lake at least this distance. This will increase the cost of the center, but does not appear to have much influence on the economic feasibility of the center.

Entrapment and Entrainment

Other possible adverse effects of an energy center on the local ecosystem include the entrapment and entrainment of organisms within the cooling system and thermal plume. The Federal Water Pollution Control Act requires that construction and capacity of cooling water intake structures reflect the best available technology for minimizing adverse environmental impact. To this end the center must demonstrate to the Michigan Water Resources Commission that the plant's operation will not seriously affect the water resources or the balance of the aquatic organisms residing in or passing through the vicinity of the plant. Furthermore, the assessment of impact must consider the cumulative effect on the water body. The principal impacts that could occur from the operation of cooling water intakes would be the damage or destruction of benthic, planktonic and/or nektonic organisms.

The major impact of an energy center with once-through cooling would arise from the large flows of circulating waters through the lake. In 30 years of operation, it has been estimated that the center at Harbor Beach would displace through its system about one-tenth of the total volume of Lake Huron. Movement of such large quantities of water could have a significant effect on lake circulation patterns. The required intake would displace through the system the equivalent of about one square mile of water 100 feet deep each day of operation. If the flow were constrained to a velocity of only 0.5 ft/sec, the intake structure area would have to be 60,000 ft^2. Such a structure would entrain a significant portion of the organisms and influence the circulation patterns of Lake Huron (see Figure 31).

To avoid a significant impact on the ecosystems of the Great Lakes, the energy center systems must be designed to minimize entrainment and entrapment, using such possibilities as:

1. A network of intake and discharge pipes such that the effects are dispersed rather than concentrated.
2. Dike a nearshore area to provide a large single circulation and cooling basin. Such a basin could also serve as a port facility for the center.
3. Construct a large radiator or heat exchanger in the lake that would utilize the lake for cooling benefits but that would provide a system free of entrainment problems.[1]

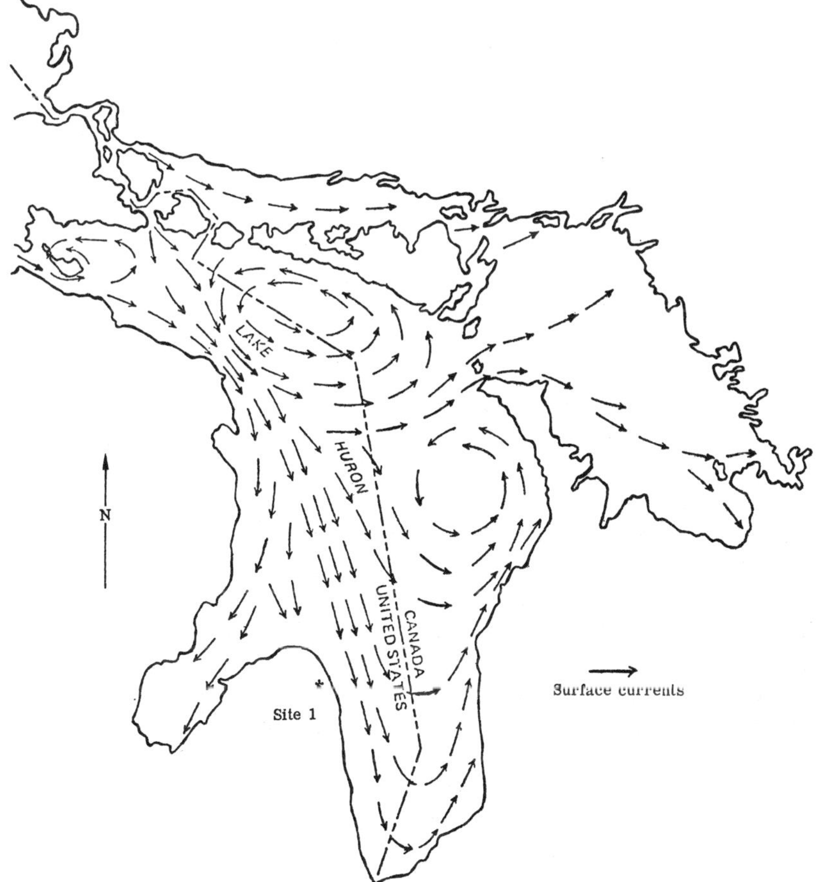

Figure 31. Surface currents, summer period, Lake Huron.[44]

These particular suggestions could either be too expensive or otherwise impractical, but some approach of this magnitude may be required if the energy centers are to fully utilize the Great Lakes' cooling capacity. The enormity of the entrainment problem again suggests a compromise using some once-through cooling utilizing the lakes' heat capacity along with cooling towers and ponds.

THE IMPACT ON AIR QUALITY

The potential environmental effects of the energy center on local and regional air quality appear to be a serious constraint for center site development. The concern for air quality includes the possible adverse effects of releasing large amounts of waste heat and water vapor into the atmosphere from ponds, cooling towers and spray channels associated with the power generation. There are, in addition, the important effects from stack emissions of fossil-fueled power plants and industrial plants. This section presents a review of the possible adverse effects of center operation on air quality and a review of existing air quality standards and controls.

Thermal Discharge

One method of waste heat disposal from energy center power plants is the employment of cooling towers. These towers transfer waste heat to the atmosphere by both evaporation and convection. The predominant heat transfer to the atmosphere is accomplished by the first process through the condensation of water vapor produced by the waste heat of the power plant. If a large number of cooling towers are crowded into a small part of the site, they can produce the so-called "heat island" effect. This effect is characterized by the formation of a stationary localized thermal plume over the site. It has been suggested that the presence of this plume might affect the local climate but the extent has not yet been determined. There will also be some increase in humidity close to and downwind of the site which may produce local fog. These effects can be reduced if the facilities are distributed over the entire site rather than concentrated.

During the operation of the cooling towers, a small portion of the water will be carried out of the tower with the saturated water vapor plume. These small droplets, or drift, have the same chemical composition as the circulating cooling water. Some concern has been expressed that the droplets will fall to the ground and the chemicals contained in the drift will be introduced into the local terrestrial ecology. Care taken to maintain the quality of the circulating water would reduce or eliminate the adverse effects of this drift. Moreover, the cooling towers can be constructed with proper drift eliminators and should produce neither additional rainfall nor icing from drift. In the operating experiences of the United Kingdom and in the U.S. these adverse effects have not occurred.

The most notable feature of the cooling tower operation is the visible plume. The operation of cooling towers could produce some ground fog in the vicinity of the towers. Usually, the plume buoyancy will produce a rise or "lift-off" within several meters of the tower and the problems will be limited to a small area relatively close to the tower. Nevertheless, the location of towers relative to one another and to the surrounding area should be evaluated over the range of weather conditions at potential sites to ensure that the effects will be minimized.

Other possible effects of using cooling towers include the noise emitted as large quantities of air and water move through the tower's internal configuration, and the potential pollution of groundwater if algicides and fungicides used inside the tower are released or if the cooling tower plumes interact and mix with the stack pollutants, such as sulfur dioxide released from refineries and fossil-fueled plants.

Atmospheric Effects from Waste Heat and Water Vapor

At this time, the effects of heat and water vapor on the atmosphere are not well known and no thorough study has been made of possible environmental consequences. It would be wise to move with caution on many fronts because of the possible undesirable meteorological effects from dissipating large amounts of waste heat and water vapor into the atmosphere. Of course any such effect could be greatly magnified at an energy center with its greater concentration of facilities. Present Michigan and federal air quality laws do not limit the air discharge of heat and water vapor as long as such a discharge does not induce a chemical reaction with other pollutants. One concern would be the possibility of oxidizing sulfur dioxide in the presence of water vapor to form sulfuric acid, which could then be precipitated in rain.

Of all the cooling methods used by power plants, cooling towers have the greatest effect on the atmosphere, since the energy flux from the top of a tower is the highest. Potential problems that might arise from the use of power plant cooling towers include the following[1]:

1. restriction of sunlight caused by a visible plume;
2. deposition of chemicals in cooling water onto surrounding areas;
3. ground fogging;
4. spatial and temporal changes in the pattern of precipitation; and
5. initiation of severe weather, such as thunderstorms.

One way to assess the effects of cooling towers on the atmosphere is to compare their energy productions and energy densities with other sources and atmospheric processes. Table XLI provides a comparison of heat production and flux among several natural processes and manmade structures. Note that the total heat flux from the waste heat of power plants compares with that from major metropolitan areas. However, the heat flux for an enterprise as large as the energy center will be a few times larger than most other anthropogenic sources. Possible meteorological effects of large releases of heat into the atmosphere may be estimated by comparison with the effects of natural disasters as shown in Table XLII. The energy center will compare in its effect to a large brush fire or

Table XLI. Energy Production Rates of Natural Atmospheric Processes and Anthropogenic Sources[45-47]

Process	Area (km^2)	Energy Production $(W\ m^{-2})$	Fraction of Solar Flux at Ground
Natural Sources			
Solar Energy Flux at Ground	–	1.6×10^2	1
Cyclone Latent Heat Release			
(Assume: half-life = 3 days			
rainfall rate = 1 cm/day)	10^6	2.0×10^2	1.3
Great Lakes Snow Squall Latent Heat Release			
(Assume: snowfall rate = 4 cm/hr)	10^4	10^3	6
Thunderstorm Latent Heat Release			
(Assume: half-life = 30 min			
rainfall rate = 1 cm/30 min)	10^2	5.0×10^3	30
Tornado: Kinetic Energy			
(Assume: half-life = 10 min)	10^4	10^4	60
Anthropogenic Sources			
Anthropogenic Heat from Cities			
Manhattan	59.8	630	3.94
Moscow	878	127	0.79
Washington, DC	173	44	0.28
Los Angeles Basin	10,000	7.5	0.05
Boston-Washington Metropolitan Area			
(projection for 2000 AD)	31,200	36	0.23
Waste Heat from Power Plants			
Summit, Salem, Hope Creek			
(12 mi-5 mi)	155	73.8	0.46
Peach Bottom, Fulton, Summit, Salem, Hope Creek, Bainbridge, Conowingo			
(50 mi-10 mi)	1,294	22.1	0.14
Typical Cooling Pool (assumed production)			
10^3 kW/2 ac	–	1.3×10^2	0.81
Typical Cooling Tower			
10^6 kW/10^4 m^2	–	10^5	600
Michigan Energy Center	38	10^3	6

Table XLII. Effects of Large Heat Additions to the Atmosphere[47]

Phenomenon	Energy Rate (MW)	Area (km^2)	Energy Flux Density ($MW\ km^{-2}$)	Meteorological Consequences
Large Brush Fire	100,000	50	2,000	Relatively small energy flux rate, very large area. Cumulus cloud reaching to a height of 6 km formed over 1/10 area of fire. Convergence of winds into a fire area.
Forest Fire Whirlwind	—	—	—	Typical whirlwind: central tube visible by whirling smoke and debris. Diameters few feet to several hundred feet. Heights few feet to 4,000 ft. Debris picked up: logs up to 30 in. diameter, 30 ft long.
Surtsey Volcano	100,000	1	100,000	Permanent cloud extending to heights of 5-9 km. Continuous sharp thunder and lightning, visible 115 km away. (Phenomenon probably peculiar to volcano cloud with many small ash particles.) Waterspouts resulting from in-draft at cloud base, caused by rising buoyant cloud.
Surtsey Volcano	200,000	<1	200,000	Whirlwinds (waterspouts and tornadoes) are prevalent. Often there is at least one vortex downwind: short inverted cones or long sinuous horizontal vortices that curve back into the cloud and intense vortices that extend to the ocean surface.
Single Large Cooling Tower	2,250	0.0046	484,000	Unknown
Array of Large Cooling Towers (Nuclear Park)	72,000	4	18,000	Unknown
Michigan Energy Center	44,000	38	1,100	Unknown

forest fire and will include cloud formation and convergent winds. It is estimated that the total heat output from an energy center will be comparable to that of a fairly large city. Some weather modification effects have been attributed to the urban "heat islands" formed over Chicago and New York. For example, the urban release of waste heat does raise the ambient temperature by several degrees to heights of a few thousand feet. The flux density of heat released by an energy center will be much greater since the total area is much smaller and the sources are discrete. The waste heat from cooling towers is likely to penetrate much higher altitudes than that from urban areas. As a rule, the more waste energy entering a given mass of ambient air, the greater the meteorological effect.

While for urban areas there is an increase in temperature only, an energy center could cause both a change in temperature and moisture. The effect might be more like that associated with lake evaporation and heat transfer to a cold atmosphere. It has been concluded that under certain conditions an energy center could increase cloudiness and rainfall and in some cases trigger thunderstorm activity. However, accurate estimates of the magnitude of this climatological impact cannot be based on information available at this time.

Chemical and Particulate Emission

In addition to the thermal discharge to the environment from power plants, the fossil-fueled plants, coal gasification and liquefaction facilities and industry can cause particulate and gaseous pollution of the atmosphere. This section discusses the types and amounts of pollutants expected from the various elements of an energy center plus a discussion of air quality standards and the likelihood of meeting these standards.

Air Quality Standards

There is a potential for large amounts of air pollutants to be discharged by fossil-fueled power plants and industries at any energy center. Table XLIII lists the types and present annual emission rates of air pollutants resulting from the operation of a typical modern 1,000-MWe fossil-fueled conventional power plant. The major sources of these emissions are the chemical constituents of the fuel utilized in the combustion process. The combined effect of all center sources may be of sufficient magnitude to create an air quality problem outside the center.

EPA ambient air and water quality standards, together with provisions of the Clean Air Act of 1970, must be met by new and existing fossil-fueled plants with respect to limits on concentration levels and effluent discharges. All industries will be required to meet the new stationary-source stack emission standards for sulfur-containing fuels as outlined in Table XLIV. It is not clear whether air quality problems will be avoided by meeting these standards throughout the area around the site. The federal ambient standards, which provide for public protection, are summarized in Table XLV. The primary standard is established to protect the public health, while the stricter secondary standard is established to protect health and welfare and prevent damage to plants, animals and the impairment of human vision. The stationary source standards can be associated

Table XLIII. Air Pollutant Emissions from a Typical 1,000 MWe Convensional Plant[48,a]

Pollutant	Annual Release (10^3 lb)		
	Coal[b]	Oil[c]	Gas[d]
Particulates–Fly Ash	9.9	1.6	1.02
Oxides of Sulfur	306.0	116.0	0.027
Oxides of Nitrogen	46.0	47.8	26.6
Carbon Monoxide	0.460	0.0184	Negligible
Hydrocarbons	1.150	1.47	Negligible

[a]Based on normal average heat rates, load factors and fuel properties.
[b]Burning 2.3×10^4 over 10 yr. Assuming 3.5% sulfur content, of which 15% remains in the ash, and a 9% ash content with 97.5% fly ash removal efficiency.
[c]Burning 460×10^4 gal/yr. Assuming 1.6% sulfur content and 0.05% ash content.
[d]Burning 68×10^4 mcf/yr.

Table XLIV. Sulfur in Fuel Limitations for Fuel-Burning Equipment

Plant Capacity 1,000 lb Steam/hr	Maximum Sulfur Content in Fuel Percent by Weight	
	July 1, 1975	July 1, 1978
0.500	2.0	1.5
Over 0.500	1.5	1.0

Equivalent Emission Rates

Sulfur in Fuels (%)	Parts Per Million by Volume Corrected to 50% Excess Air		Pounds of Sulfur Dioxide per Million Btu of Heat Input	
	Solid Fuel (12,000 Btu/lb)	Liquid Fuel (18,000 Btu/lb)	Solid Fuel (12,000 Btu/lb)	Liquid Fuel (18,000 Btu/lb)
1.0	590	490	1.8	1.1
1.5	800	600	2.4	1.7
2.0	1,180	840	3.2	2.2

Table XLV. Federal Ambient Air Quality Standards

	Primary	Secondary
Suspended Particulates (μg/m^3)		
Annual geometric mean	75	
Maximum 24-hr concentration[a]	260	150
Sulfur Oxides (μg/m^3)		
Annual arithmetic average	80 (0.03 ppm)	
Maximum 24-hr concentration[a]	365 (0.14 ppm)	
Maximum 3-hr concentration[a]		1,300 (0.5 ppm)
Carbon Monoxide (μg/m^3)		
Maximum 8-hr concentration[a]	10 (9.0 ppm)	10
Maximum 1-hr concentration[a]	40 (35 ppm)	40
Photochemical Oxidants (μg/m^3)		
Maximum 1-hr concentration[a]	160 (0.08 ppm)	160
Nitrogen Oxides (μg/m^3)		
Annual arithmetic average	100 (0.05 ppm)	100
Hydrocarbons (μg/m^3)		
Maximum 4-hr concentration[a]	160 (0.24 ppm)	160
(6-9 am)		

[a]Not to be exceeded more than once a year per site.

with air quality conditions existing in a large urban area. The EPA air quality implementation plans call for classification fo all areas of the U.S. according to regional air quality conditions. Each class has a proposed limit on allowable increases in total suspended particulates and sulfur dioxide in order to prevent significant deterioration of existing air quality.

The federal government published on December 5, 1974, the following description of air quality classes: "Class I applied to areas in which practically any change in air quality would be considered significant; Class II applied to areas in which deterioration normally accompanying moderate well-controlled growth would be considered insignificant; and Class III applied to those areas in which deterioration up to the national standard would be considered insignificant. Under the proposed regulation, all areas of the country would be designated Class II initially with provisions for allowing States to reclassify any area to accommodate the social, economic and environmental needs and desires of the public."

For areas designated Class I and Class II, the deterioration is to be limited to increases over the baseline air quality concentrations as follows:

	Class I	Class II
Particulates, 24-hr max.	10 μg/m^3	30 μg/m^3
Sulfur Dioxide		
24-hr max.	5	100
3-hr max.	25	700
Annual Average	2	15

If the EPA implementation plan is put in operation, large portions of Michigan would be unsuitable for an energy center since it will be designated Class I. These areas include large portions of the Upper Peninsula and the northern Lower Peninsula. The rest of the state will likely be designated Class II, with the exception of the Detroit Metropolitan area.

EPA has divided the state of Michigan into six Air Quality Control Regions (AQCR), as shown in Figure 32. Those neighboring counties with similar topography and industry are grouped into a single region. Continuous monitoring of air quality is being conducted in each AQCR by the Michigan Air Pollution Control Division. Sulfur dioxide concentrations for the counties of the central AQCR, which include the Michigan study sites, are given in Table XLVI. In Huron County at the Harbor Beach Site, the SO$_2$ values are dominated by emissions from the Hercules Powder Company. A value of around 40 μg/m^3 rather than 91 μg/m^3 would be a more realistic value for Huron County as a whole. With this value, an energy center would be allowed to deteriorate the ambient air quality to 140 μg/m^3, or to less than one-half the national standard.

There will be some sites where the energy center will be subjected to ambient standards that are more stringent than the national standards. In this case special precautions would have to be taken in planning the center layout. These precautions may include appropriate separation of stack emissions to reduce the potential additive effects from multiple sources, an appropriate placement of the stacks with respect to prevailing winds to avoid a cumulative effect, and other design arrangements. In the past, the effect of multiple sources has been determined simply by summing emissions from contributing concentrated sources. For a system as large as an energy center this may result in an overly stringent requirement that could be relaxed somewhat by careful design. There are several models available to calculate concentration levels of plumes for each contaminant making use of typical meteorological conditions at the proposed site.

Factors Contributing to the Spread of Effluents

The most important factors in the spread of power plant and industrial effluents are the local meteorology, climate and hydrology. The most

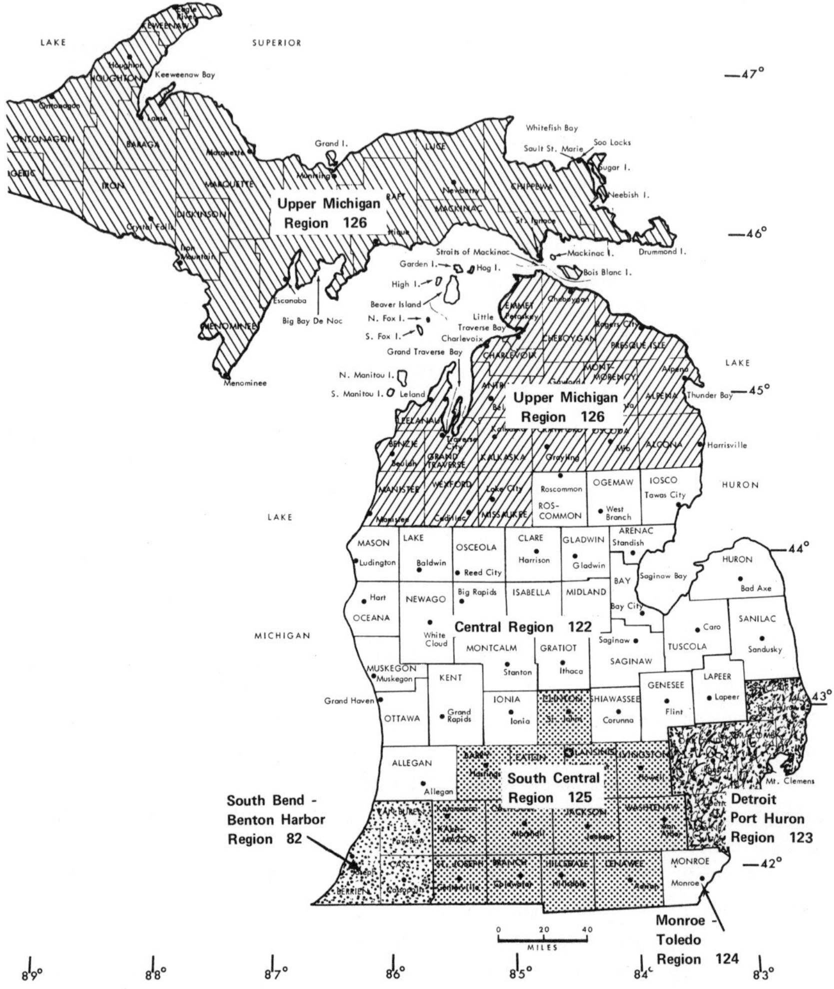

Figure 32. Air quality control regions.

important considerations of meteorology and climate are:

1. the configuration of the local air flow;
2. dilution of effluent by the movement of the local air mass, which is dependent on the average wind speed, the prevailing wind direction, the variability of wind conditions, atmospheric turbulence, the verticle structure of the temperature, prevalence of temperature inversions and the roughness of local terrain;
3. various removal processes, such as chemical processes in the atmosphere,

gravitational precipitation, washout from precipitation and condensation processes and, in the case of nuclear effluents, radioactive decay; and

4. ambient temperature and humidity.

Table XLVI. Summary of Sulfur Dioxide Levels Monitored in the
Central Michigan AQCR

County	Second Highest 24-hr Value (μg/m^3)	Maximum Annual Average (μg/m^3)
Bay	440	46
Genesee	185	32
Gratiot	265	47
Huron	550	91
Kent	125	45
Midland	680	89
Muskegon	1,529	118
Ottawa	375	96
Saginaw	205	24

Hydrologic factors influencing the effects from the spread of effluents include: (1) the number and types of local bodies of water; (2) the amount of surface drainage; (3) the amount of subsurface water; and (4) the local geologic structure. To determine the amount and effect of effluent spread at a given site, use is made of several existing models. These models require detailed data for the site as listed above. In some cases, these data may already be available, but usually large-scale data-gaterhing operations at the potential site are required. EPA requires that air and water quality measurements be performed in the vicinity of potential power plant and industrial sites, usually by the state's Department of Health. The sources of particulate matter include industrial and agricultural activities, incineration and road dust. In conjunction with the construction of a new coal-fired plant, a utility should collect background ambient air quality data for sulfur and nitrogen oxides to compare with the input of the plant when it is operating.

The Problem of Sulfur Dioxide Emission Levels

For coal-fired power plants, the emission of sulfur dioxide is probably more important than any other emitted particulate. Although emission control technology is available for removing sulfur dioxide, such as scrubbers, fuel desulfurization and tall stacks, there are still uncertainties as to whether these approaches can meet EPA requirements. This section will examine whether the combined fossil-fueled power plants and associated industry will exceed the allowable limits of SO_2 emissions and, consequently, the ambient air quality standards. The ground-level measurements of SO_2 will depend upon the stack height, exhaust flow, mixing height, prevailing wind speed and direction, and the area's stability class. Several dispersion models exist to predict these effects but their validity has yet to be evaluated. The Michigan energy center study team made use of the previous study made by General Electric.[13] The appropriate scaling was used to make the results comparable to the scope of the Michigan energy center. The GE study provides a simulation of short- and long-term concentrations of sulfur dioxide, which can be used to predict the potential air quality problems of the Michigan center. The assumptions and characteristics of the GE study are as follows:

1. Fossil-fueled power plant capacity: 1,320 MWe; stack height, 800 ft; exhaust velocity, 46.5 ft/sec.
2. Use EPA recommendation for the maximum 15-min ground-level concentration of SO_2.
3. Assume energy center consists of 12 point sources; each point equivalent to two 1,320-MWe generating units.
4. Assume constant prevailing winds so as to give rise to highest concentrations.
5. No optimization to minimize ground-level concentrations through stack placement.
6. No consideration is given to use of intermittent controls such as low-sulfur coal or part-time load operation.

In order to use the results of the GE study for the Michigan case it was necessary to assume that the local meteorological conditions for Rochester, New York, are representative of that occurring at either of the Michigan energy center sites. This may not be a bad assumption since the predominant wind characteristics are similar. The sulfur dioxides emissions from eight 885-MWe coal-fired generating units have been estimated from the General Electric study as follows:

$$28,500 \text{ ton/yr of } SO_2 \text{ for one 1,320-MWe unit}$$

$$8 \times \frac{885}{1,320} \times 28,500 = 153,000 \text{ ton/yr}$$

In evaluating the potential ground-level concentrations, comparisons are made with ambient and proposed Class II stnadards. The estimated cumulative probability distributions of ground-level concentrations for 3-hr and 24-hr standards are shown in Figures 33 through 35 for an energy center similar to that in the Michigan study. With a system to provide for 80% removal of sulfur dioxide from the stack emissions, the Michigan energy center is estimated to meet the Class II 3-hr standard of 700 $\mu g/m^3$. This has been calculated as if the sources have been confined to a small area. If appropriate attention is paid to dispersing the sources, the ground-level emissions would be somewhat less.

Examination of Figure 34 indicates that the Class II and national ambient air quality 24-hr standards will be exceeded about 1% and 7.5% of the time, respectively. According to these curves, an 82% removal of SO_2 from the energy center stack emissions would be required to meet the primary standard, while 87% removal would be required to meet the Class II maximum deterioration standard of 100 $\mu g/m^3$. If the energy center stacks are concentrated in a 0.84-mi^2 center, then 80% removal of SO_2 will meet the 24-hr concentration standards all but 27 days per year. If the stack sources are spread out over the proposed 25-mi^2 energy center, the situation would be improved. In the Michigan energy center study, the sources are assumed to be located in the four corners of the 5 x 5-mi area. Each corner site contains two fossil-fueled plants and collocated industry. The plumes generated by this arrangement would

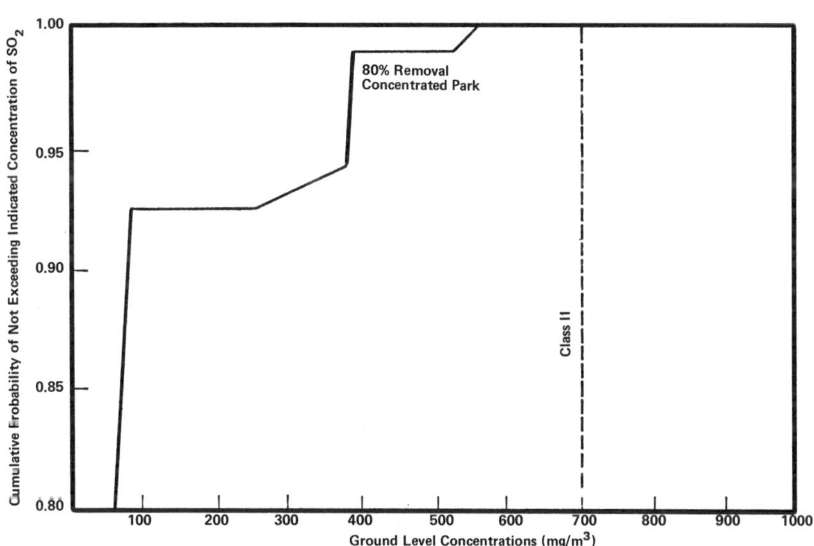

Figure 33. Cumulative distribution of 3-hr concentration.

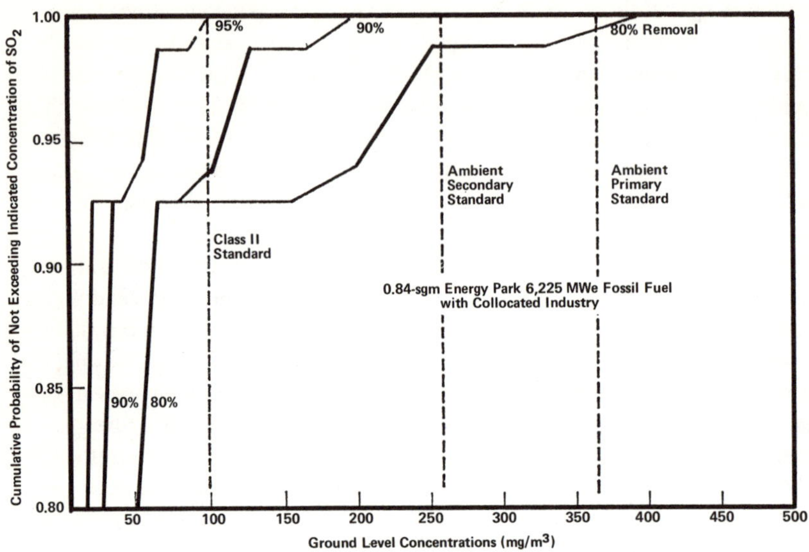

Figure 34. Cumulative distribution of 24-hr concentration.

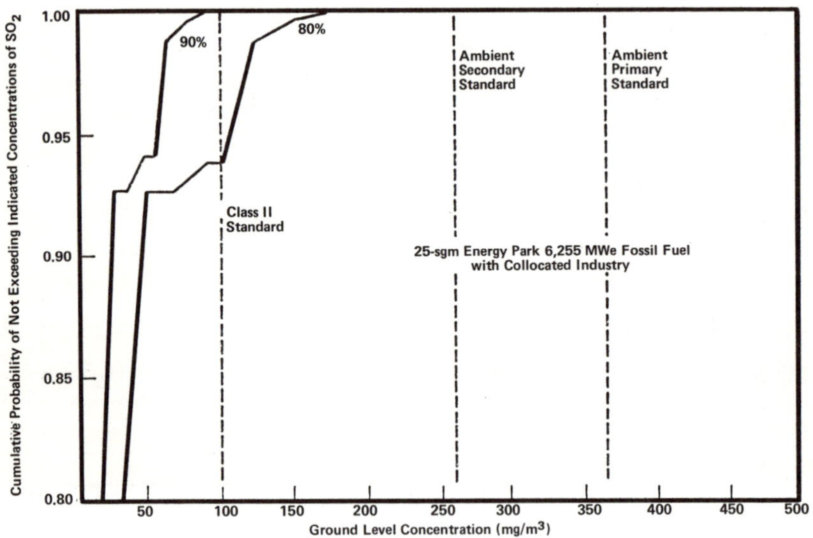

Figure 35. Cumulative distribution of 29-hr concentration.

result in additive effects, but the maximum ground-level concentrations occur only near each corner. In this case the national ambient standards can easily met with 80% SO_2 removal. For this case, the Class II standard would be exceeded about 7% of the time and 90% removal would be required for the spread-out center to meet the Class II standard all the time. It was estimated that the maximum annual average ground-level concentrations of the Michigan energy center would be only half the Class standard of 15 $\mu g/m^3$.

The Michigan energy center study team concluded that if the system meets the stack emission standards it will not have a problem meeting the national ambient standards nor the Class II standards. One possible exception might be the 24-hr maximum sulfur dioxide level for Class II areas, which would be exceeded about 23 days per year. This figure may be reduced both by greater dispersal of the stack units or by reducing electrical power output during certain critical days, or by taking measures to remove more than 90% of the SO_2 emissions. If all these approaches are impractical, the currently conceived Michigan energy center could not be constructed in a Class II area.

NUCLEAR EFFLUENTS

The fission processes of a nuclear power plant create radioactive products with a variety of chemical properties. These products decay with half-lives ranging from fractions of a second to around 30 years. During normal operation, very small quantities of these fission products may be released from small defects in the cladding of the nuclear fuel elements. Trace quantities of radioactive materials are emitted to the atmosphere through various leaks in the nuclear reactor and its containment building or are discharged to local receiving waters through the cooling cycle. These so-called routine emissions always exist to a certain extent no matter how secure the storage facilities. The quantities and character of the emissions vary depending on the reactor type and the specific design of the reactor. Typical emissions from a boiling-water reactor are listed in Table XLVII as reported by the U.S. Public Health Service. The units of radioactivity are given in curies.

The radioactive materials released from nuclear plants are dispersed in the environment and transported along several pathways which may result in exposure to man. The most important pathways by which radionuclides from a nuclear facility may reach the local population by atmospheric discharges are[1]:

1. whole body external exposure;
2. inhalation exposure;

Table XLVII. Typical Emissions from One Boiling-Water Reactor

| Isotope | Emission Rate | | | Quantity in Existence $(Ci)^{a,b}$ |
	Half-Life	Ci/sec	Ci/yr^a	
Noble Gases				
^{83m}Kr	1.86 hr	240	7,600	2
^{83}Kr	10.76 yr	0.6	20	300
^{85m}Kr	4.4 hr	400	13,000	9
^{87}Kr	76.0 min	1,300	40,000	8
^{88}Kr	2.80 hr	1,400	44,000	20
^{89}Kr	3.18 min	150	4,700	0.04
^{133}Xe	5.27 day	400	13,000	300
^{135}Xe	9.14 hr	1,400	44,000	70
^{135m}Xe	15.6 min	1,000	32,000	1
^{137}Xe	3.9 min	350	11,000	0.1
^{138}Xe	17.5 min	3,000	90,000	5
Others				
^{3}H	12.262 yr	6×10^3	0.2	3
^{58}Co	71.3 x	2.6×10^{-5}	8×10^{-4}	2×10^{-1}
^{60}Co	5.263 yr	2.5×10^{-5}	8×10^{-4}	6×10^3
^{89}Sr	52.7 day	9.7×10^{-4}	0.03	6×10^{-3}
^{90}Sr	27.7 yr	5×10^{-6}	2×10^{-4}	6×10^{-3}
^{131}I	8.05 day	9.2×10^{-4}	0.03	9×10^{-1}
^{137}Cs	30.0 yr	3.5×10^{-5}	1×10^{-3}	5×10^{-2}
^{140}Ba	12.80 day	4×10^{-4}	0.01	9×10^{-1}

[a]Assuming full-time operation through the year.
[b]After operation time with respect to the half-life.

   3. deposition on grass → animals → man
   4. deposition on soil or plants → man.

The pathways through liquid discharges are:

   1. waterway → drinking water supply → man
   2. waterway → fish, etc. → man
   3. waterway → aquatic plants → animals → man
   4. waterway → external exposure.

At any given energy center, the major concern is the number of different pathways by which radioactive effluents can travel to man. The number of pathways and the ease of conduction are each dependent on the site selected.

In addition to the number and nature of effluent pathways to man, another important factor determining radiation exposure is the population

density and distribution around the nuclear facility. Regulations of the
Nuclear Regulatory Commission require that the area immediately surrounding a site must be divided into zones of restricted population such that
individuals on the borders receive doses of radiation, as a result of normal
operation of the plant, that are within certain low limits. These regulations
prescribe an exclusion zone devoid of population, then a low-population
area, and further a requirement that the distance to a population center
(containing more than 25,000 residents) be at least one and one-third
times the distance from the reactor to the outer boundary of the low-
population zone. The critical factors with regard to site safety are the
surrounding population densities and distributions and the local meteorological and hydrological conditions. These factors determine the spread of
radioactivity in the air and water and its quantity and extent of exposure
to man.

The maximum limit of radiation dose at the exclusion-area boundary
coupled with local atmospheric dispersion characteristics will determine
the distance to the exclusion boundary and, therefore, the size and shape
of the energy center site. Given these considerations, the different possible
exposure pathways, the total population and its distribution, the reactor
type, and the physical environment of the area, it is possible to estimate
the total dose for any given individual. For example, if it is assumed that
radioactive liquids are discharged into a river, then the persons most
likely to be exposed to radiation would be those who consume fish raised
near the outfall, drink water drawn from the river, boat near the outfall,
and/or swim and conduct shoreline activities within one mile of the outfall. Similarly, if radioactive gases from a facility are released at ground
level and dispersed in the area by diffusion, the individuals most likely to
receive the highest dose would be those who live near the site's exclusion
boundary in the direction of the least favorable dispersion and those who
consume leafy vegetables and milk produced near the site.

Environmental surveillance at nuclear facilities to evaluate critical
exposure pathways is necessary but difficult. Established NRC regulations
include monitoring gases and airborne particles prior to release from exits
into the atmosphere and monitoring liquid waste before discharge into the
coolant water and prior to dilution with the receiving water. Other environmental samples are routinely analyzed, such as local milk, for the
presence of radionuclides that may have originated in the reactor. Better
estimates of radiation dose to man can be obtained from actual environmental measurements, but it is often more practical to estimate exposures
from release rates and the assumed exposure pathways. These estimates
may be performed by the use of the several available analytic or computer
models of dosage as a function of the characteristics of the nuclear

facility and the surrounding area. To determine these estimates, the models use fission-product release rates, local atmospheric dispersion characteristics and local population density structure. Each critical pathway is considered separately in calculating doses for a specific radionuclide. Common types of estimated individual exposures include doses to an adult's total body or to various organs. Individual exposure is described in units of rads or rems and a useful concept for evaluating radiation dose to human population is the total "man-rem." This is defined as the product of the exposure to manmade radiation of individuals in the population group times the number of people exposed.

Maximum concentrations of radionuclides as well as individual and population doses at the exclusion-area boundary are regulated by the Nuclear Regulatory Commission and as recommended by the International Commission of Radiation Protection (ICRP) and its U.S. counterparts, the National Council on Radiation Protection and Measurements and the Federal Radiation Council. The ICRP-recommended limits on maximum permissible concentrations and doses at the exclusion-area boundary are provided in Table XLVIII for various critical organs and fission products.

It will be noticed from Table XLVIII that the maximum permissible dosage for various organs of the body vary from 12 to 30 rems per year. The previous Table XLVII also gives the radioactive effluents from a typical power plant. The values are not commensurate since the units are different; one is given in rate of discharge while the other is given in maximum dosage. Two approaches are helpful in evaluating the emissions from nuclear power plants. First, no nuclear power plant operating today is allowed to give off a radiation dose in excess of 5 millirems per year to any individual living near the plant boundary. This is very much less than the permissible levels indicated in Table XLVIII. Second, it is informative to compare these levels with other sources of radiation. The routine radioactive emissions from a normally operating nuclear power plant are much less than the normal background radiation of 150-200 millirems (although it is additive). The increase radiation of a nuclear plant (on a yearly exposure basis) can be compared to one round-trip air flight from New York to Los Angeles (4 millirems). Operating experience of present nuclear power plants has shown that routine emissions can be held to much less than the 5-millirem requirement; therefore these effluents present little or no risk.

Since the routine emissions of nuclear power plants result in an addition to the natural background emission and, therefore, could be undesirable for that reason, it is very important to compare nuclear power effluents with the alternatives. One recent study by Cuddihy *et al.* [51] has performed a simulation study of the radiation risks from plutonium recycle.

Table XLVIII. ICRP-Recommended Limits on Maximum Concentrations
and Doses at the Exclusion-Area Boundary[49]

Isotope	Maximum Permissible Concentration (μCi/ml)	Critical Organ	Maximum Permissible Dose to Organ (rem/yr)
H-3	0.03	Whole Body	12
Sr-89	1.0×10^{-4}	Bone	30
Sr-90	1.0×10^{-5}	Bone	30
Y-90	2.0×10^{-4}	GI	15
Y-91	3.0×10^{-4}	GI	15
Zr-95	6.0×10^{-4}	GI	15
Nb-95	1.0×10^{-3}	GI	15
Mo-99	2.0×10^{-3}	Kidney	15
I-131	2.0×10^{-3}	Thyroid	30
I-133	7.0×10^{-5}	Thyroid	30
Te-132	3.0×10^{-5}	GI	15
Cs-134	9.0×10^{-4}	Whole Body	12
Cs-136	9.0×10^{-5}	Whole Body	12
Cs-137	2.0×10^{-4}	Whole Body	12
Ba-140	3.0×10^{-4}	GI	12
La-140	2.0×10^{-4}	GI	15
Ce-144	1.0×10^{-4}	GI	15
Mn-54	1.0×10^{-3}	GI	15
Co-58	0.03	GI	15
Co-60	5.0×10^{-4}	GI	15
Fe-59	6.0×10^{-4}	GI	15
Cr-51	0.02	GI	15

They concluded that "predictions of health-effects models, by themselves, cannot justify an 'acceptable' or 'insignificant' level of disease incidence in the exposed population. Greater meaning can be gained by using the same population exposure methods to compare similar health hazards from alternative energy industries."

The magnitudes of radioactive effluents from a pulverized coal-fired power plant were reported by Kaakinen *et al.*[52] "After passage through mechanical dust collectors and an electrostatic precipitator, the effluent contained 3.51 mCi of ^{226}Ra, 166 mCi of ^{210}Po, and 150 mCi of ^{210}Pb per year for a 1,000-MWe plant. . . . Population risks for exposure to these effluents of coal power plants were also simulated. This analysis projected incidences of 5.4×10^{-2} lung tumors, 1.6 bone tumors, 3×10^{-3} liver tumors and 1.86 genetic defects in the exposed population for each year of operation of a 1,000-MWe coal plant. Compared to the

operation of an equivalent size LMFBR, this was 400 times the incidence
of lung tumors, 15,000 times the incidence of bone tumors, 150 times
the incidence of liver tumors and 21,000 times the frequency of genetic
defects." These estimates, if anywhere near accurate, provide a sobering
comparison between the health dangers from coal-fired and nuclear plants.
"The greater numbers of projected health effects from coal burning
result from larger releases of alpha-emitting radionuclides, their higher
gastrointestinal absorption and their greater deposition in internal organs
after absorption. Uptake of transuranic radionuclides by plants was
assumed to result in a plant-to-soil concentration ratio of 0.1 and the
same factor was used for lead, polonium and radium. This was considered
to overestimate plant absoprtion of all of these elements."

"Alpha-emitting radionuclides in effluents from power generators are
only one risk to which human populations are exposed. There are also
hazards from mining operations, from transport of materials and a variety
of mechanical operations as well as from handling of by-products and
wastes that may be radioactive or chemically toxic." Because of the much
larger quantities of fuel used in coal-fired plants, many of these additional
hazards are greater than for nuclear plants.

Abnormal Operations

Routine radioactive emissions appear to be no risk for normally operating nuclear plants. Of much greater concern is the possibility and conse-
quence of large radioactive release from the occurrence of an abnormal
external or internal event at a nuclear reactor. The risks of large releases
from abnormal occurrences have been analyzed and reported by
Rasmussen.[40] By NRC regulations, nuclear reactors are designed to withstand such accidents as floods, earthquakes, tornadoes and impacts by
cars, aircraft and windblown objects, such as utility poles. The Rasmussen
report and others have suggested that the probability of a large release
of radioactivity as a result of such an accident is on the order of only
10^{-6} per reactor-year.

Internal abnormal occurrences are defined as nine classes of reactor
accidents which are related to the failure of various reactor components
and safety systems. The higher the class of accident, the greater the potential release of radioactivity—ranging from amounts that are small compared
to routine operational releases to catastrophic releases of large fractions of
the total radioactive inventory—and the lower the estimated probability of
occurrence. The appropriate design of the nuclear containment vessel
would preclude the release of the entire radioactivity inventory and, therefore, safety analyses consider that the worst possible accident that can

occur results from a sequence of failures that lead to a core meltdown and release of a large amount of radioactivity. For the core-meltdown type of accident, the risks and consequences of each possible type of accident are evaluated in terms of the dosage of various radionuclides to an individual at the exclusion-area boundary. The probability of a core meltdown has been estimated to be around 10^{-5} per reactor-year. At an energy center, the probabilities of a core-meltdown accident involving more than one reactor unit would both have very small probabilities. Nevertheless, the safety problem for the series of reactors at an energy center may be more severe since one core meltdown may well require area evacuation of the entire area. This could place the other units in jeopardy or require the shutdown of all the other nuclear units. In an energy center, there is a need to determine the safety implications of the possible interaction between units; these analyses have not yet been performed in detail.

Combination of Fossil and Nuclear Effluent Effects

At an energy center with both fossil- and nuclear-fueled power plants, it is necessary to consider not only additive but also synergistic effects, that is, an interaction of the effects of the individual effluents which leads to an enhancement of results. Such possible synergistic effects have not been studied in detail and their magnitude and implications at an energy center have not been determined. Yet, several possible considerations relating to possible synergistic effects might be predicted:

1. interactions between cooling tower plumes and the chemical and nuclear effluents emitted by the fossil-fueled and nuclear power plants, associated industries, and stored fuel and waste deposits;
2. interactions among components of the increased amount and variety of chemical air and water pollution emitted by the entire center; and
3. interactions among the components of the totality of emitted chemical and nuclear effluents.

The full implications of these synergistic effects are not yet clear, however, some effects can be anticipated. For example, cooling tower plumes may upset the normal area micrometeorology and change the local atmospheric dispersion from its predicted behavior. This could allow the spread of effluents in unpredictable ways and present increased hazards to local populations and the land. In addition, some combinations of pollutants, such as certain particulates and sulfur dioxide, can have effects that are more hazardous to people's health than the sum of their independent effects. And finally, the effects of health and crops from the combined action of chemical and radioactive air pollutants is largely unknown and could be severe.

INFLUENCE OF SITE CONDITIONS ON
HEALTH AND SAFETY FACTORS

Many natural disasters, such as floods, earthquakes, tornadoes and accidents, might effect the health and safety implications of an energy center. The history of former occurrences at the site serves as a predictor of the probability of future similar events. Energy centers are best located in areas free of dams or unusually large precipitation. Such events could produce damage to site facilities which would aggravate the consequences of polluting liquid or gaseous effluents released to the environment.

Another site-related selection factor arises from the remote possibility of an aircraft crash at power plant locations. Clearly, the probability decreases with distance from an airport. Yet the concentration of industrial facilities would require airports to be within practical distances of the center. In addition, the concentration of facilities may increase the severity of external accidents, as well as of internal accidents such as explosions, fires or accidental hazardous chemical releases.

Health and Safety Implications of
Siting Energy Centers in Michigan

The detailed analysis of the climatological data presented in an earlier chapter suggests that relatively benign weather and the absence of severe climatic conditions favors the Huron County site. Atmospheric stability and wind structure in this region provide favorable dispersion characteristics for vapor plumes and airborne contaminants arising from the center. Power transmission facilities of a center are most vulnerable to severe icing occurrences and tornadoes. Since an energy center will require clustering of several transmission lines at one location, damage from a single storm occurrence could seriously hamper power reliability. However, the Huron County site is at the extreme edge of the Midwest tornado belt and the probability estimate of a tornado striking any one point in this area is only 5×10^{-4} per year. Since 1900, only one tornado is known to have touched down in Huron County. This low tornado activity is viewed as an advantage for center siting in this area compared to more southerly locations. Freezing rain and icing conditions can be expected for approximately 50 hours. This is relatively light and favors the Harbor Beach location.

The Michigan energy study team described the Huron County area as very favorable with respect to the geological and seismic implications of power plant siting. There are no known active faults within 200 miles of Harbor Beach nor are there any known cavities within the region. The existence of good bedrock below the surface provides favorable foundation conditions. The entire state of Michigan, particularly the thumb area, has a history of very low seismic activity. Because of the sensitive siting requirements for nuclear power plants, the thumb area is highly favored over alternate sites in the Midwest and indeed the entire U.S.

THE ENVIRONMENTAL IMPACT OF OTHER ENERGY CENTERS

Not surprisingly, the environmental impacts predicted by all the energy center studies so far conducted are very much the same. Significant differences arise either from different center sizes or are site-specific. In keeping with the objectives of this book, only those differences and environmental impacts which are significant will be discussed.

In contrast to the Michigan energy center sites with their ample water from the Great Lakes, the Pennsylvania study concludes that the hydrological system for the region will be burdened with a heavy water demand. Because of the quantities of water needed, most sites in Pennsylvania would require construction of reservoirs. In this state the quantity of water removed from a river during low-flow periods is regulated. Since approximately 250 to 300 ft^3 of water per second will be evaporated by the Pennsylvania park of 10,000 MWe, no Pennsylvania river could meet this demand without a reservoir, even if 10% of the 10-yr low flow is the restricted consumption rate. A reservoir will be required to supply cooling water and maintain minimum river volumes during low-flow periods. The size of this water reserve will depend critically on both effective low-flow regulations and low-flow expectations for the river.[3]

The water restrictions for the Puerto Rico energy center are somewhat eased by the possibility of using seawater for cooling purposes. Nevertheless, this center will require freshwater for process purposes. The irrigation water requirements for the Guayama Subregion are largely satisfied by the existing South Coast Irrigation District and water pumped from wells. It is expected that the demand for irrigation water will either remain steady or decline in future years as food crops other than sugar cane are introduced which require less water, and as sprinkler irrigation, that uses water more efficiently than the present furrow-type irrigation, is put into use.

Existing industrial and municipal wells produce 7.8 mgd at the Puerto Rico site; this could be supplemented with 4.0 mgd from wells on the West Aguirre Site presently used for irrigation. The Toa Vaca Dam project is scheduled to deliver 28 mgd in the near future and 75 mgd in 1989 which will meet a major part of the center requirements. The quality of this water is expected to be adequate for direct use in most industrial processes. For municipal purposes, further treatment would be needed. Since these additional sources will require time to develop, the water shortages must be supplied either from other wells of from desalting plants. The Puerto Rico plan anticipates the use of desalting plants to make up this deficit. In keeping with the symbiotic advantages of energy centers, the Puerto Rico study team expects to utilize desalting plants not only for the production of freshwater, but also for providing salt to feed the chlorine-caustic plant which would use 3,000 tons of salt per day.

A review of the plans for all the energy center studies to date shows that there will be a significant, if not crucial, impact on the availability and allocation of current water supplies. With the exception of the Michigan sites, there will no doubt be a requirement for evaluating trade-offs of available water uses.

The atmospheric pollution and radiation dose impacts of the various centers have been judged to be about the same, as would be expected since the impact figures were taken to be representative of the current state-of-the-art. For example, it was concluded by the Oklahoma study team that the numerous facilities could be sited at Camp Gruber without creating radiation doses in excess of the current design objectives and without increasing the average radiation dose to the surrounding population by more than 0.01 percent. Similar conclusions were drawn for the Puerto Rico and Wasatch Front energy centers given appropriate weighting to size differences. The impact on atmospheric pollution from coal-burning power units will depend on the efforts to remove stack pollutants, most notably SO_2. These efforts, in turn, will affect the capital costs required and subsequently the costs of electricity and the economic feasibility of the center. Early estimates from all studies suggest that air pollution can be held to acceptable limits and that energy centers can be feasible. However, center feasibility is very sensitive to the costs of environmental control and detailed economic studies must still be performed. Nevertheless, an energy center does not substantially alter the impacts resulting from generation of the required amounts of electric power; the impacts tend to be larger yet more easily contained because of the concentration of facilities.

XII

SUMMARY:
ARE ENERGY CENTERS THE CITIES OF THE FUTURE?

In previous chapters, the advantages and problem areas of energy centers have been discussed. It is the purpose of this chapter to assess whether the overall advantages outweigh the problems. Since all energy center studies to date have been preliminary, the conclusions must of necessity be tentative. The conclusions of the individual study teams will be used to attempt this assessment.

CONCLUSIONS FOR THE MICHIGAN
ENERGY CENTERS

A major change brought about by the construction and operation of an energy center at either Michigan site will result from the influx of construction workers, operating personnel, managers, their families and other people providing required services for the immediate area. The labor force at the Michigan center will reach a peak of about 13,000 people. Considerably more employment will be required to provide commercial and government services to these employees and their families. There will be a major impact on the existing population in the area, particularly to the farm families and other residents around Harbor Beach or Muskegon.

The rapid birth of a new city would occur too fast to allow normal growth evolution to provide adequate housing and required services. Unplanned growth, with its real estate speculation, zoning inadequacies, poor temporary housing and urban blight, is undesirable. Growth rates, similar to those expected at Michigan energy centers, have been accomplished with success elsewhere; with proper planning, the rise of a new city can be achieved advantageously. A total planned community, similar to Columbia, Maryland, or other successful attempts, should be considered a part of the center planning.

The major environmental impacts that should be guarded against by center design and monitoring are expected to be:

1. the undetermined effects from the discharge of large amounts of waste heat to the atmosphere from cooling towers and cooling ponds;
2. the possible effects on water quality from discharging inadequately treated process water and from groundwater recharge and storm runoff with center-contaminated waters;
3. the dispersion of large amounts of sulfur dioxide and other pollutants that may exceed Class II standards; and
4. the release of radioactive effluents during normal operations or as a result of abnormal external or internal events.

The problem of obtaining financing may be a major deterrent to center construction. The financing of any energy center complex will require very large amounts of capital, which is difficult to obtain in today's financial markets. It should be emphasized, however, that the problem of providing sufficient capital will occur whether or not energy generation is to be concentrated at an energy center. On the other hand, financing the center will be especially sensitive to this concentration and the usual methods used by local private utility companies may not be adequate. Special treatment through government participation by tax incentives, subsidies for essential public services, and furnishing or guaranteeing funds needed for center development may have to be considered. In order to assure the stability of essential financing, whatever the source, reasonable assurance is required that legal, political, environmental and other impediments to rapid and economical center construction will not be excessive.

The Michigan energy center study team[1] concluded that their study uncovered no severe impediments to the construction and operation of energy centers at locations similar to the study sites at Harbor Beach and near Muskegon. It was recognized that most of the problems and negative impacts resulting from an energy center will also be realized from alternative approaches to meeting the required power demand. An energy center is attractive because it enables these problems and impacts to be solved in a controlled and orderly manner at a single location by a single organization. The study team emphasized that there are difficult problems inherent in an energy center concept, but the nature and size of the energy center which give rise to these new problems also provide the built-in mechanisms for their solution. The ability to achieve the desired economies and efficiencies contained in the concept implies a need for a single agency with sufficient power and authority to solve those problems.

The Michigan study did not directly compare other alternatives to the energy center. It was the team's opinion that the data strongly suggested that most of the advantages inherent in economies-of-scale can still be maintained while making better use of process team and waste heat by splitting the energy generation facilities into two or more centers with their own collocated industries and biocomplexes. One major finding in the Michigan study was that only 12% of the waste heat could be utilized in a biocomplex associated with a 24,000-MWe generating facility. This limit would be approached because it is not economical to transport warm water long distances and, therefore, agricultural applications must be located very close to the generating units. Since agriculture uses a great deal of land, only that land adjacent to the generating facility is close enough for economical transport of the warm water. On the other hand, splitting the generating facilities into two or more smaller centers enables more farmland to be within transport distances.

A similar situation would limit the magnitude of collocated industries which can utilize process team. High-temperature high-pressure steam can be economically transported only a few miles, thus limiting its use to industries located quite close to the power-generating facilities. Again, the land available that meets this criterion is limited and only a few adjacent industrial concerns can make use of the steam. Note that an energy center that expands to a fourfold increase in area would have only a twofold increase in its perimeter, where the collocated industry must be situated.

Another factor which favors two or more centers arises from the potential problems of meeting environmental regulations. A 24,000-MWe center, the number of nuclear reactors assumed by the study team, would exceed the current limit allowed at a single site. Additionally, the coal-fired units would exceed the Clean Air Act limitations for emissions about 23 days out of a year. Both regulations could be met at two or more smaller centers. The Michigan study concluded that it may be more desirable to locate 16,000 MWe at Harbor Beach and 8,000 MWe at the western site. Such a split would more closely match the existing load center requirements and would allow shorter and less expensive transmission lines.

With the split into two smaller centers, the overall conclusion of the Michigan energy center study team was favorable. The problems, though difficult, could be solved with careful planning, and there would be a 3 5% decrease in capital investments required and 12% or more of the waste heat could be economically utilized. Collocation of industry and district heating and cooling of the associated center city would achieve the much desired conservation of additional energy.

CONCLUSIONS OF THE CAMP GRUBER
ENERGY CENTER STUDY

An important aspect of the preliminary study of the feasibility of siting a large complex of energy facilities at Camp Gruber was estimating the impact of the center on the surrounding region. Camp Gruber[2] appears to be a satisfactory site for many types of energy facilities. The study team concludes that the size of the complex would probably be limited primarily by the amount of high-quality water available for use. "Most energy facilities probably could be sited at Camp Gruber with minimal impact because of the Camp's favorable environmental characteristics and because the facilities would have to be built in accordance with all applicable environmental, health and safety standards. Based on a preliminary survey, most local residents appear to accept what they perceive to be the environmental and socioeconomic impact of the energy center. Since a portion of Camp Gruber is used by the public for recreational activities such as hunting, plans for recreational activities should include the objective of minimizing interference with these activities.

"The study identified two topics as needing further investigation and planning: (1) Obtaining enough water for an energy center. This could be in the form of a comprehensive water supply and allocation plan for eastern Oklahoma. The plan should include a way of supplying the energy center with an adequate amount of water. (2) Preparing an energy center development plan. Establishing an energy center and reaping the maximum benefits with minimum environmental impact calls for an administrative plan for its development. The plan for Camp Gruber should include a determination of the best administrative structure for promotion and development of the center, and performance of research necessary to decide the types of facilities best suited to the Camp.

"Whether or not an energy center *ought* to be sited at Camp Gruber was not the subject of [their] study. That question will, presumably, be the subject of future study and discussion within the State of Oklahoma."[2]

CONCLUSIONS OF THE COMMONWEALTH OF
PENNSYLVANIA ENERGY CENTER STUDY

The study team determined that the energy park alternative is not likely to reduce the consumer's cost of electricity by any significant amount. While the energy center may require more land than the dispersed siting alternative, the environmental effects of the center would not differ qualitatively from dispersed sites, though the impact would be more intense on the site. It was decided that the center's generating units will place substantial demands on the involved watershed. It was also decided that the social and economic impact would be less than previously forecast by studies within the Commonwealth.

CONCLUSIONS OF THE PUERTO RICO ENERGY CENTER STUDY

It was concluded that an excellent opportunity exists for government and industry to work together to accomplish their separate ends, namely, that of government to create employment and that of industry to engage in profitable business. The primary physical assets that Puerto Rico offers are a desirable plant site with a deep water harbor and a work force to build and operate the energy center. No native raw materials would be used.

The major contributions that Puerto Rico offers are governmental policies that encourage industrialization through tax exemptions and quotas, and the ability to raise money by tax-free bonds for the construction of certain facilities. The Puerto Rico study team concluded that the center could be implemented through the existing agencies of the government of Puerto Rico and that plans can be formulated for organizing the governmental-industry relationship to realize the economic benefits of the energy center as projected.

CONCLUSIONS OF THE WASATCH FRONT AREA ENERGY CENTER STUDY

After performing a first-phase study, it was concluded that an industrial chemical complex could be constructed near Salt Lake City which would utilize Utah raw materials, provide additional employment, furnish an industrial base for further development and yield adequate profits.

APPENDIX

LEGAL AND REGULATORY FACTORS IMPACTING ON ENERGY CENTERS IN MICHIGAN

FEDERAL REGULATORY STATUTES AND THE ANTITRUST LAWS

This section deals with the general applicability of federal laws and regulations to energy centers.[1] It is the objective of the discussion to indicate some of the legal factors that should be taken into consideration when specific center configurations, organizations, financing arrangements, etc., are evaluated.

The major part of the material on this subject was abstracted from a variety of reports, memoranda and legal analyses prepared for a number of private industrial firms and government agencies. In particular, Cravath, Swaine, and Moore; Upton and Doane; and Truitt and Fabrikant were major sources of information.

Federal statutes that could have significant impact on the financial structure and operations of energy centers in the state of Michigan include The Federal Power Act, The Public Utility Holding Company Act, The Sherman Antitrust Act, The Clayton Antitrust Act and the Robinson-Patman Act. The specific applicability of the laws will vary, depending on the form of the energy center and whether or not the power generated at the center is sold and/or transmitted over state lines. Thus, an industrial firm that incident to its process steam requirements generates power to sell to a utility might be dealt with differently than a power generation facility that is the result of a joint financial venture between a utility and an industrial firm, and which enters into long-term power supply agreements with both the utility and the industrial firm. In a similar sense, a separate utility that supplies power to other privately or publicly owned industrial or agricultural industries collocated on the energy site might be dealt with differently than either of these two configurations.

241

The Federal Power Act

The provisions of the Federal Power Act and the jurisdiction of the Federal Power Commission (FPC) apply to the transmission of electric energy in interstate commerce and to the sale of electric energy at wholesale in interstate commerce [16 U.S.C. 824b (1970)]. The act specifically exempts the sale or transmission of energy in intrastate commerce, and does not deprive any state of its lawful authority over the exportation of hydroelectric energy over a state line.* The jurisdiction of the FPC, when the act applies, extends over facilities for the transmission and sale of electricity, but not over facilities for the generation of electric energy. The act defines the term "sale of electrical energy at wholesale" to mean the sale of electric energy to any person for resale [16 U.S.C. 824d (1970)], and the term "public utility" to mean any person who owns or operates facilities subject to the jurisdiction of the FPC [16 U.S.C. 824e (1970)].

When an industrial firm is held to be a public utility within the meaning of the act, remaining provisions of the act authorize the FPC to exercise various authorities [16 U.S.C. 824 & 825 (1970)]. Among these are authorities over the physical connection of transmission facilities of the public utility with other engaged in interstate transmission and sale of electric energy; the sale, deposition or merger of facilities; the issuance of securities**; the acquisition of securities of another public utility; unjust or discriminatory rates and charges; adequacy of service; maintenance of depreciation accounts; the filing of reports; the records and books of the public utility; and the holding of office by directors who are also directors of other public utilities, banks and investment banking firms. The regulations of the FPC also authorize an annual assessment against each public utility to defray the costs of administration of the act.

In general, if the facilities are not involved in the transmission or sale of electric energy in interstate commerce, the act will not apply to the facilities. However, where a transaction involving the facilities *affects* interstate commerce, it may be found to fall under federal jurisdiction and the authority of the FPC [*United States v. Yellow Cab Company*, 332 U.S. 218 (1947); *Jersey Central Power and Light Company v. FPC*, 319 U.S. 61 (1943)]. Thus, even assuming that the Michigan energy center power generation facility is designed only for intrastate distribution of energy, as has been postulated by the study team, the requirements placed on the energy center power facility under various kinds of pooling arrangements could result in the transmission of power in interstate

---

*A Michigan energy center configuration could contain pumping stations utilizing natural bodies of water. Thus, hydroelectric regulation could pertain to the energy centers.

**Authority over the issuance of securities by the public utility does not extend to utilities organized in states in which security issues are regulated under law by a state commission.

commerce and thereby bring the facility under the authority of the FPC.*
For example, the Michigan Power Pool provides for a central dispatch
center, and the companies which are parties to the agreement have agreed
to meet reliability criteria for their facilities to prevent power interruption.
In addition to power pools, Michigan utilities are included in the Mid-
America Interpool Network (covering the western section of the Upper
Peninsula), and the East Central Area Reliability Coordination Agreement
(which covers the rest of the state). Thus, should any of the parties to
these agreements become the operator or owner of the power facility in
the energy center and make the facility a part of the pool or other
regional agreement, and should any energy from any of the utilities find
its way across the state line, the FPC could exercise jurisdiction over the
energy center facility.

There are unresolved questions about whether or not a power generation facility organizationally and/or financially separated from a transmission facility is subject to FPC jurisdiction when some of its power finds
its way across a state line. Some case law [*Connecticut Light & Power
Co. v. FPC*, 324 U.S. 515 (1945); hearings on Senate Bill 1725, 74th
Congress, 1st Session (1935)] and the legislative history of The Federal
Power Act appears to support the view that a generation facility which
is not part of a utility venture is not subject to the jurisdiction of the
FPC. However, it is possible that the venture could become subject to the
jurisdiction of the FPC if it sold power to a public utility. Since the FPC
has rate review authority over public utilities, with respect to both prices
paid and the prices charged for electric power, the rates charged by the
generating facility for power sold to the utility might be subject to review
by the FPC.

As was stated earlier, once FPC jurisdiction is established over the
power facilities, a large number of authorities can be exercised by the FPC.
One area of interest that might be subject to review by the FPC are long-
term power sales contracts with collocated industries at the energy center
site. For a completely balanced facility, it would appear that long-term
commitments would have to be made to the associated industrial/agricul-
tural facilities by the power facilities. In cases involving the Securities and
Exchange Commission (SEC), a long-term power purchase contract was
treated as a guarantee, or an assumption of liability on securities issued
by the corporation selling the power, and thus a security of the purchaser
[*Wisconsin River Power Company et al.*, 27 SEC 539 (1948)]. Accordingly,
the question is raised whether or not a long-term power purchase contract
might be treated as a security by the FPC if the purchasing party is subject to FPC jurisdiction, since the act requires authorization of the FPC
for the assumption of any obligation, or liability as a guarantor of the
security of another (unless the issuance of utility securities is regulated

---

*In Michigan, there exists a Michigan Electric Coordination Agreement between Con-
sumers Power Co. and the Detroit Edison Co., Consumers Power Co. Rate Schedule
No. 33 (1973), filed with the FPC.

under law by a state commission). It is generally believed that where the long-term contract is not a prerequisite to the financing of the project (energy center), it will not be viewed as a security of indebtedness issued by the purchaser. However, if the agreement calls for payment for power whether the power is taken or not, and the power purchase prices are designed to meet the operating expenses of the power facility, the answer to the question becomes less clear. Under the interpretation that the security is an indebtedness and the rules of the FPC, such a long-term contract might have to become the subject of competitive bidding, unless an exemption is secured from the FPC.

The Public Utility Holding Company Act

The Public Utility Holding Company Act subjects public utility holding companies, their subsidiaries and their affiliates to the jurisdiction of the Securities and Exchange Commission (SEC). A "holding company" is defined as any company that directly or indirectly owns, controls or holds with power to vote, 10 percent or more of the outstanding voting securities of a public utility company or of a company which is a holding company; or any person who directly or indirectly exercises such a controlling influence over the management or policies of any public utility or holding company that it is appropriate in the public interest or for the protection of investors or consumers for the person to be subject to the act. However, the SEC, upon application of the person or company, can declare such person or company not a holding company, if the Commission finds that the person or company does not meet the jurisdictional descriptions given above.

The Commission can also declare a company not a public utility if the Commission finds that the company is primarily engaged in one or more businesses other than that of an electric utility company and sells so little electric energy that it is not necessary in the public interest to consider such a company not a public utility. It can also find a company not a public utility if the company operates within a single state and substantially all of its outstanding securities are owned by another company to which the subject company sells or furnishes electric energy for its own manufacturing use.

It is of interest to examine what is meant by the term "intrastate." Case law has stated that the operations of a utility are predominantly intrastate when all of the property of the utility is within the state and the utility makes no substantial sales outside of the state or at the state line of out-of-state consumers [*Monarch Mills*, 1 SEC 822 (1936) and *Texas Utilities Co.*, 31 SEC 367 (1950)]. Unfortunately, there are insufficient SEC decisions to determine the level of sales or revenues that can be used as a guide to what are "substantial" sales.

The results of becoming subject to the jurisdiction of the SEC are as follows: the holding company must register with the SEC, the sale and issuance of securities must conform to SEC regulations, and the acquisition of securities and utility assets of other businesses require the approval of the SEC. Jurisdiction also provides the SEC with authority to limit and simplify the operations of the holding company, to require reports be submitted to the SEC, to require the maintenance of a uniform set of accounts, and to prevent certain forms of interlocking directorates and officers.

The Sherman and Clayton Acts

The Sherman and Clayton Antitrust Acts are designed to prevent combinations of business and industry that can act in restraint of trade among the states. The Sherman Act provides that every contract, combination in the form of a trust or otherwise, or conspiracy in restraint of trade or commerce among the several states, or with foreign nations, is illegal; and provides that every person who monopolizes, or attempts to monopolize, combines or conspires with others, to monopolize any part of trade or commerce among the states is guilty of a misdemeanor. The Clayton Act provides that no corporation engaged in commerce shall acquire, either directly or indirectly, the whole or any part of the stock or other share capital of another corporation also in commerce, where the effect of such acquisitions would be to substantially lessen competition or tend to create a monopoly.

Some problems with respect to the antitrust laws might result from an energy center if only one electric utility has access to the energy generated by the power facility and if that power is significantly cheaper than could be obtained by competitors. The same holds true if the industrial or agricultural enterprises in the center have the financial benefits of the combination to the detriment of their respective competitors. However, since the exclusion of other utilities or industrial/agricultural enterprises from access to the power arrangement would have a valid basis other than combination for the purpose of injuring competitors, it is not likely that such problems would actually arise. For example, it is particularly significant that the only feasible arrangement for a utility's buying power from the energy center would have to include guarantees that a minimum amount of uninterruptable power would be available to the utility for an extended period of time. This requirement could result in an exclusive arrangement for power between that utility and the power generation facility and prevent other utilities from purchasing power. Moreover, no problem with such arrangements should arise since the arrangement between the power generation facility and the associated industries and/or utilities will not in any way prevent the construction of other generating facilities to feed other utilities and industries.

Additional antitrust problems might arise with regard to process steam and waste heat, which by the nature of the energy center arrangement would be limited to those industrial/agricultural enterprises collocated on the center site. Since these associated activities would have to be planned so as to optimize the use of the limited process steam and waste heat available, other enterprises would be precluded from obtaining any benefit from the power source. The practicalities of the matter appear to preclude problems in this area, however, since no industry and/or agricultural enterprise would enter into the arrangement unless it was assured an un-interruptible source of such heat energy; thus, the arrangement would have a valid basis rather than one designed to injure competitive activities.

Other problems that might arise would be concerned with domination of an area with respect to power generation if two competing utilities entered into an agreement to contract and operate the power generation facility and to divide the available power markets between them. (It appears that such a division of markets authorized by the State Public Utility Commission would not pose any problems.) With respect to this problem, the specific nature of planned center arrangements would require detailed analyses so as to preclude any arrangements that would pose such problems.

As a general rule, it can be argued that any cooperative arrangements among the power generation facilities, utilities and/or associated industries making up the energy center would have a valid and legitimate purpose, and would not be designed to provide a vehicle whereby competitors or potential competitors entered an agreement to divide market, fix prices or stifle competition. Thus, it is believed if the arrangement is carefully drawn, many of the potential antitrust problems can be avoided.

The Robinson-Patman Act

The Robinson-Patman Act has as one of its primary purposes the avoidance of price discrimination in interstate commerce where such dis-criminations would tend to lessen competition or create monopoly. It does not prohibit price differentials that are a result of the differing costs that arise from differences in quantities manufactured, mode of delivery, etc. Thus power, process steam and waste heat from a single generation facility would have to be sold at the same price to all associated indus-trial/agricultural facilities on the site, subject to the cost differences arising from operating factors such as those indicated.

REGULATORY PATTERNS IN MICHIGAN[1]

The United States Constitution grants Congress the power to regulate commerce among the states and declares that the law promulgated under this constitutional grant shall be the supreme law of the land (U.S. Con-stitution, Article VI). Thus, state statutes and regulations that conflict with federal statutes are invalid [*Illinois Natural Gas Co. v. Central Illinois Public Service Co.*, 314 U.S. 498 (1942)] , and those aspects of the gen-eration and transmission of electrical energy and other products moving in interstate commerce that are specifically covered by federal statutes cannot be regulated by the states. Where specific legislation reserves the regulatory function to the states, as in the preservation of state authority over the issuance and sale of public utility securities in The Federal Power Act, a state may legislate in the area. Thus, public utility security areas, as well as generation facilities, local distribution facilities and new power plant, construction will generally be left to individual state devices and agencies.

In Michigan, jurisdiction over utilities is exercised by the Michigan Public Service Commission (PSC). It is vested with complete power and jurisdiction to regulate all public utilities in the state, except municipally owned utilities and others restricted by law [Michigan Compiled Laws Annotated, 460-6 (1967)]. A "public utility" in Michigan is defined to include persons and corporations, other than municipal corporations, or their leasees, trustees and receivers owning or operating equipment or facilities in the state for producing, generating, transmitting, delivering or furnishing electricity for the production of light, heat or power to or for the public for compensation [Michigan Compiled Laws Annotated, 460.502 (1967)]. These definitions have been interpreted to include only companies owning or operating facilities devoted or dedicated to "public use"; that is, the product or service generated by the facilities are available to the general public without discrimination. Depending on the form of the venture in the energy center, a power-generating facility might or might not be classified as a public utility; the classification being determined by whether the facility served the general public or a select group of consumers [*Michigan Consolidated Gas Co. v. Sohio Petroleum Co.*, 321 Mich. 102, 32 N.W. 2d 353 (1948)].

Once jurisdiction is exercised over the public utility, the PSC has authority to review utility rates and practices to insure that the public pays no more than a just and reasonable price and the utility receives a just and reasonable return on its investment. In Michigan, the PSC is required to investigate complaints received about unjust rates and, upon finding that the rates are unjust or unreasonable, to issue orders modifying the rates. It should be noted that the PSC has no authority to review rates under Federal jurisdiction for electric energy sold wholesale in interstate commerce.

In addition to providing a fair return on investment to the public utility, just and reasonable rates must also be nondiscriminatory [Michigan Compiled Laws Annotated, 460.557 (1967)]. However, the signing of long-term power purchasing agreements does not appear to violate this nondiscriminatory requirement, since the reduced cost that might result from such an agreement in an energy center situation should be viewed as the purchaser's capital commitment to the generation of electrical power rather than any special rate adopted by the public utility.

The PSC also has the authority to regulate the sale and issuance of securities [Michigan Compiled Laws Annotated, 460.301 (1967)]. Michigan law does not require securities to be sold at competitive bid. The

PSC also has jurisdiction over plant location and construction, requiring that a certificate of public convenience and necessity be issued before any facility can be constructed and/or operated [Michigan Compiled Laws Annotated, 460.502 (1967)]. With this authority, the PSC can effectively control power generation and transmission facility locations and service areas so as to avoid overlap of service in the state. This authority could have significant impact on any planned energy center in the state.

The authority of the PSC appears to be limited to transmission and distribution of electric power rather than generation of such power. The law specifically states that when electricity is generated, transmitted and delivered to the consumer, only the transmission and distribution, and the rates, rules and conditions of service relating to that business, *i.e.*, transmission and distribution, are subject to regulation. Similar limited authority appears to hold over safety, reliability and adequacy of service; that is, the Commission apparently is limited to jurisdiction over transmission lines installed in public places [Michigan Compiled Laws Annotated, 460.551, 460.552, 460.555, 460.556 (1967)]. The generation of electric energy appears not to be covered. The PSC also has the authority to require the maintenance of uniform accounts and records, and the submission of periodic reports to the PSC by the utility companies.

ENVIRONMENTAL PROTECTION REGULATIONS

Both federal and state environmental laws will have an impact on energy centers in Michigan. These laws and the regulations that are promulgated under them will apply to any energy center configuration, independent of corporate or organizational structure, whether or not they constitute utilities.

Federal Environmental Protection Statutes

The federal statutes that are applicable to energy centers in Michigan include The National Environmental Policy Act of 1969 [42 U.S.C. 4321-47 (1970)], The Clean Air Act [42 U.S.C. 1857-58 (1970 ed. Supp. I [1971])], The Federal Water Pollution Control Act [33 U.S.C. 1151-75 (1970 ed. Supp. II [1971])], The Coastal Zone Management Act of 1972 [16 U.S.C. 1451-1464 (1970)], and The Marine Protection, Research and Sanctuaries Act of 1972 (PL 92-532). In addition, the Nuclear Regulatory Commission (NRC) licenses nuclear plants in the United States.

The National Environmental Policy Act (NEPA)

The NEPA has as its purpose the establishment of a national policy that will encourage productive and enjoyable harmony between man and his environment, promote efforts that will prevent or eliminate damage to the environment and biosphere, and stimulate the health and welfare of man [42 U.S.C. 4332 (2)(c)(1970)].

To accomplish its purpose, NEPA requires all federal agencies to include with every recommendation, report on proposed legislation or other major federal actions that will significantly affect the quality of the human environment, a detailed statement about (1) the environmental impact of the proposed action, (2) any adverse environmental effects that cannot be avoided if the proposal is implemented, (3) any alternatives to the proposed action, (4) the relationship between local short-term uses of man's environments and the maintenance and enhancement of long-term productivity, and (5) any irreversible and irretrievable commitments of resources that might be involved in the proposed action is implemented [42 U.S.C. 4332 (2)(c)(1970)].

A "major" federal action has been defined as one that requires substantial planning, time, resources or expenditures, and an action "significantly affecting the environment" has been defined as an action that directly or indirectly has an important and meaningful effect upon any of the facets of man's environment [*Citizens Organized to Defend the Environment v. Volpe*, 353 F. Supp. 520 (S.D. Ohio 1972)]. It is necessary that the action be both "major" and have a "significant environmental effect" before an impact statement is required. ("Major action" and "significant environmental effect" are separate and independent standards [*Hanley v. Mitchel*, 460 F. 2d 640 (2d Cir. 1972)].)

The term "major" implies that certain federal actions are "minor" and do not warrant an impact statement [*Jultus v. City of Cedar Rapids, Iowa*, 349 F. Supp. 88 (N.D. Iowa 1972)]. However, federal projects involving large expenditures and projects to which the federal government contributes large amounts of money have been held to require impact statements [*Environmental Defense Fund v. Corps of Engineers*, 325 F. Supp. 728 (D.E.D. Ariz. 1971); *San Antonio Conservation Society, et al. v. Texas Highway Dept.*, 446 F. 2d 1013 (5th Cir. 1971)]. Major federal actions also have been held to include federal licensing [*Green County Planning Board v. FPC*, 455 F. 2d (2d Cir. 1973)], approval of transmission lines under The Federal Power Act [*Calvert Cliffs' Coordinating Committee v. US AEC*, 449 F. 2d 1109 (DC Cir. 1971)], and approval of railroad tariffs [*Students Challenging Regulatory Agency Procedures v. U.S.*, 346 F. Supp. 189 (D.D.C. 1972)].

It is not clear whether FPC action, assuming that the energy center came under the jurisdiction of the FPC, would constitute sufficient action to justify the issuance of an impact statement. Whether or not an impact statement would be required would depend on the specific nature of the venture organized to build and operate the energy center facilities. However, as indicated in *Students Challenging Regulatory Agency Procedures*, FPC regulations generally do not require the preparation of impact statements for review of electric rates.

Any SEC actions with regard to energy centers do not appear to constitute major federal actions, since the SEC has jurisdiction over neither the construction nor the operation of energy centers.

The Clean Air Act

The Clean Air Act has as its purpose the protection and enhancement of the quality of the nation's air resources so as to promote the public health and welfare and the productive capacity of the population [42 U.S.C. 1857 (b)(1)(1970)]. Under the act, the Administrator of the Environmental Protection Agency (EPA) is required to publish a list of air pollutants that have an adverse effect on public health and welfare and that result from mobile or stationary sources [42 U.S.C. 1857-3(a) (1970)]. For each pollutant for which criteria have been issued, the Administrator must prescribe a "national primary ambient standard," whose attainment and maintenance is required to protect the public health and which must be achieved and maintained by each State through a State implementation plan [42 U.S.C. 1857c-4 (1970)].

The act also required the Administrator to formulate a list of categories of stationary sources that contribute significantly to air pollution and to establish emission standards for new sources in such categories that reflect the degree of emission limitation that is achievable through application of the best system of emission reduction [42 U.S.C. 1857c-5 (1970)].

The act defines "new source" as any stationary source whose construction or modification is begun after the publication of regulations prescribing a standard of performance applicable to such a source [42 U.S.C. 1857c-6(a)(2)(1970)]. A "modification" is defined as any physical change, or change in the manner of operation that increases the amount of any air pollutant emitted, or results in the emission of any air pollutant not previously emitted [42 U.S.C. 1857c-6(a)(4)(1970)].

It appears that an energy center that is located where no stationary sources are currently located would constitute a new sources that the Administrator would be required to promulgate emission standards to which the center would have to conform. It appears that these standards should be dependent on the specific site selected.

Enforcement of air standards is performed by the EPA Administrator. However, enforcement authority can be delegated to those states whose enforcement procedures are found to be adequate by the Administrator.

In addition to setting emission standards for stationary sources, the Administrator also have authority to set emission limits for all hazardous air pollutants, and no new sources may be constructed and no existing sources may be modified unless the source, when properly operated, does not violate these standards [42 U.S.C. 1857c-7 (1970)]. Here again, the Administrator may delegate enforcement authority to the state.

The Federal Clean Air Act does not preclude state or local enforcement of emission standards or pollution abatement if such standards or limitations are not less stringent than those set under the act [42 U.S.C. 1857 (a)(3)(1970)].

Federal Water Pollution Control Act

The Federal Water Pollution Control Act was originally passed in 1948. The 1965 amendments to the act stated that the purpose of the act was to enhance the quality and value of the water resources of the nation and to establish a national policy for the prevention, control and abatement of water pollution [33 U.S.C. 1151(a)]. The 1972 amendments to the established national goals for water quality that would (1) prohibit toxic pollutants, (2) insure the protection and propagation of fish, shellfish and wildlife, (3) see that waters are suitable for recreation by 1983, and (4) eliminate the discharge of pollutants into navigable water by 1985 [33 U.S.C. 1251(a)].

As with The Clean Air Act, the states have the primary responsibility for controlling water pollution and must submit for approval to the Administrator standards of water quality that satisfy the requirements of the act [33 U.S.C. 1151(b), 1160(c)(3)]. Once initial state water quality standards have been approved under Section 1160, they must be reviewed at least once every three years.

New sources of water pollution (including new steam electric power plants) must comply with standards of performance that reflect the greatest degree of effluent reduction achievable through the best available demonstrated control technology, processes, operating methods and other alternatives [33 U.S.C. 1316(a), (b)]. Thermal waste is included as a pollutant. New sources are defined in the same manner as in The Clean Air Act. Performance standards for generating plants became effective November 7, 1974.

The major sources of water pollution that might arise in the energy center are the cooling water used to condense the steam coming out of the turbine generator; this process can result in thermal pollution, as the water periodically is discharged ("blown down") from the condenser boiler system to remove impurities. Good design and integration of the energy center facilities would do much to eliminate the thermal problem. However, the impurity discharge problem cannot be solved so easily. Complicating the problem are the discharges that can originate with the industrial/agricultural facilities collocated in the energy center.

Under the act, electric generation facilities are permitted to discharge water at a higher temperature than the water into which the discharge is made. However, the discharge standards established under the act with respect to thermal pollution require that the location, design, construction and capacity of cooling water structures reflect the best technology for minimizing adverse environmental impact. If a thermal discharge limitation is shown by the operator or owner of a point source to be more stringent than necessary to assure a balanced, indigenous population of fish, shellfish and wildlife, the Administrator may ease the limitations applicable to the source [33 U.S.C. 1316(a), (b) and 33 U.S.C. 1326(b)].

To insure compliance with standards set under the act, permits for discharges which are effective for no more than five year, must be acquired prior to any discharge. Such permits may be issued by the state when the Administrator finds that state permit programs satisfy requirements under Section 1342(b) and 1314(h)(2) of the act. Such state programs must include programs of continuing planning for all navigable waters in the state and in annual strategy for solving state water quality problems.

The act expressly provides that no action taken by EPA, other than the issuance of permits for new sources, requires an impact statement. Since the energy center will constitute a new source, the granting of permits could constitute a federal action under NEPA, thus requiring an impact statement. However, the second requirement must also be met; the thermal discharges must have a significant impact on the environment. Thus, it appears that both The Federal Water Pollution Control Act and NEPA will play a decisive role in the processes employed to determine the feasibility of siting energy centers in Michigan.

Coastal Zone Management Act

The Coastal Zone Management Act of 1972 [16 U.S.C. 1451-1464 (1970)] encourages states to develop programs to preserve, restore and enhance the coastal zone consonant with a proper balance of competing interests. The shores of the Great Lakes are covered by this act. The act requires states to develop a plan to identify boundaries of its coast zones and develop a process to

1. determine appropriate land and water uses;
2. designate areas of particular concern;
3. establish priority uses within specific areas of the zone;
4. determine intergovernmental arrangements needed to conduct an effective management program; and
5. evaluate the adequacy of existing regulations for proper land and water use management.

Michigan is in the second year of a two-year development plan program. If the development plan meets specified criteria, the state can become eligible for implementation funds from the federal government (usually awarded on a state matching-fund basis).

Guidelines for the Coastal Zone Management Act require that power plant siting be considered a legitimate coastal zone use in the development plans. However, the Michigan Coastal Zone Management development plan may have implications for other shoreland uses related to any energy center, such as development of a commercial harbor to serve the center or use of shorelands for residential subdivisions.

Marine Protection, Research and Sanctuaries Act

The Marine Protection, Research and Sanctuaries Act of 1972 (Public Act 92-532) includes a provision to preserve or restore coastal marine areas for conservation, recreation, ecological and esthetic values through establishment of marine sanctuaries. The Marine Sanctuaries program is administered by the Office of Coastal Zone Management in the National Oceanic and Atmospheric Administration (NOAA). Major objectives of the program include designation of areas

1. to protect valuable, unique or endangered marine life;
2. to complement and enhance public areas such as parks, seashores, refuges, etc.;
3. important to the survival and preservation of the nation's fisheries and other ocean resources; and
4. to advance marine ecosystems research.

In an initial survey of potential marine sanctuaries, several were identified in the Great Lakes. One of these is Saginaw Bay, which was classified as a Priority II level on a three-level scale.

Under provisions of the act, the Secretary of Commerce "shall issue . . . regulations to control any activities permitted within the designated marine sanctuary, and no permit, license or other authorization . . . shall be valid unless . . . permitted activity is consistent with the purposes of [the act]."

While this act is not directly applicable to the proposed energy center, it could have significant ramifications pertaining to development of the Saginaw Bay shoreline for housing or other activity associated with population growth caused by the center.

Nuclear Regulatory Commission (NRC)

The NRC licensing process for nuclear plants is divided into two permit stages: the construction permit and the operating license. The NRC's formal siting regulations are limited to the specification of broad site criteria relating only to radiological health and safety consideration. These site criteria have been supplemented by specific NRC guidelines that require detailed descriptions of the environmental features of a proposed site and the environmental impact resulting from the construction and operation of a nuclear power plant on the site. NRC also requires analyses of available and practical alternative sites and types of power plants, which analysis must show that none of the alternative sites is preferable to the one proposed.

NRC issues operating licenses for fixed periods of time, not to exceed 40 years. The continuing jurisdiction of NRC over licensed activities during the license period includes detailed reporting requirements; NRC inspections of the licensee's records, premises and plant operations, and authority to require modification of operating procedures and the retrofitting of the facility. Each license contains specific conditions and detailed limitations relating to both radiological and environmental aspects of the facility operation. The NRC may also revoke, suspend, modify or amend licenses on the basis of conditions revealed by any report, record or inspection, or because of false statements by the applicant.

In addition to an operating license, the operator of a nuclear power plant must obtain other licenses authorizing the acquisition, ownership, possession and/or use of nuclear materials. If the facility contains fuel-cycle activities, other regulations of the NRC will apply. (The energy center conceived by the study team does not include collocated fuel cycle activities.) In addition, the transportation of radioactive materials, such as fuels, spent fuels and fission products, is governed by at least two agencies besides the NRC—the Department of Transportation (DOT) and the U.S. Postal Service.

Endangered Species Act

The Federal Endangered Species Act of 1973 (Public Act 93-205) and the State Endangered Species Act of 1974 (Public Act 203) apply to all lands of the state of Michigan. Under the federal act, only two specific endangered fauna species occur in Michigan—the Kirtland's warbler and the eastern timber wolf. Neither are found, nor are likely to be found, locally in the Harbor Beach or Montague study areas.

The federal endangered plant species list is still in the compilation process. The tentative list, published in the July 10, 1975, issue of the *Federal Register*, does contain a number of Michigan species, but Michigan botanists questioned the selection of species, both those included in and those excluded from the federal list. Butcher's thistle, for example, is considered endangered by Michigan botanists, but is excluded from the list. It is a species known to have been collected in Huron County.

Because of the unique position of Michigan in the transition zone between boreal and central hardwood vegetation zones, there is a likelihood that a number of flora species endemic to Michigan will be on the endangered list. The position of Huron County in the transition zone between the beech-maple region and the hemlock-white pine-northern hardwoods region of the eastern deciduous forest further indicates that unique species may be found in the Harbor Beach site area. The ancient glacial lakeshore topography of ridges and swales and dunes, typical of the "tip-of-the-thumb" shoreline area, supports a vegetation association similar to that in the northern part of the Lower Peninsula. In fact, it represents the southern-most expansion of this type of vegetation.

The Michigan Endangered Species Act was patterned after the federal act and enacted into law in 1974. Species lists, administrative rules and so on are in the process of being formulated. At this time, it seems unlikely that any endangered species plant or animal other than Butcher's thistle will be found in the Harbor Beach area. However, there may be a number of threatened species, epsecially of waterfowl and plants that would occur in the area.

The status of the federal and state acts and a more precise understanding of the flora and fauna of the area need to be ascertained prior to commitment of the area for use as an energy center. It might well be, for example, that micrometeorological changes (weather modification),even on a small scale, resulting from construction of an energy center could irrevocably alter the climate-soil-topography association that supports the ridge-swale-dune vegetation association (braken fern, polygala, clintonia and other more northerly types). This could compromise the integrity of such specific sites as the Federated Women's Club Wildflower Sanctuary south of Oak Beach in Hume Township.

Other Federal Regulatory Statutes

Although not necessarily related directly to environment, a number of other Federal statutes could have an impact on energy centers in Michigan. These acts include (1) The Rivers and Harbor Appropriation Act of 1899 [33 U.S.C. 403 (1970)], which prohibits the construction of any structure in any navigable river or waterway unless a permit is issued by the Corps of Engineers, and (2) The Hazardous Materials Transportation Act (Public Act No. 93-633, 88 Stat. 2156), which is enforced by DOT.

State Environmental Statutes

Michigan, like other states, has passed its own statues to protect the environment. Such statutes include local zoning ordinances. The discussion that follows will be limited to Michigan Statutes with statewide application, affecting air and water quality and power plant siting.

Michigan Agencies

In Michigan, divisions of the Department of Natural Resources have major responsibilities for environmental monitoring and protection. The Michigan Air Pollution Control Commission establishes ambient air standards and issues permits for the construction and operation of air pollution control facilities and source emissions [Michigan Compiled Laws Annotated 336.15 (Supp. 1974)]. The Commission is empowered to suspend enforcement of the act where local air pollution control ordinances or rules will result in substantial compliance with the act and related federal laws.

The Michigan Water Resources Commission sets pollution standards and issues permits to assure compliance with state and federal regulations pertaining to industrial and commercial discharges [Michigan Compiled Laws Annotated 323.5 (Supp. 1974)]. No discharges may be made without such a permit.

In addition to these statutes, a bill requiring certification by the Michigan Public Service Commission of proposed construction of electric generating facilities has been introduced in the Michigan Legislature (House Bill 6253). The act would apply only to companies generating electricity for the public at rates regulated by the PSC.

The Department of Public Health and Department of Agriculture also have responsibilities which would affect center construction and operation.

Environmental Protection Act

The Michigan Environmental Protection Act (MEPA) (Act No. 217, Public Acts of 1970) provides for legal actions for the protection of air, water and other natural resources and the public trust therein. Authority to maintain such actions is given to the Attorney General, any political subdivision of the state, any agency of the state, and any person, partnership, corporation, association, organization or other legal entity. Actions may be maintained against the state, any of its agencies or subdivisions, or any person or legal entity within the state.

MEPA provides citizens the right to defend themselves through the courts from all forms of environmental pollution. The act was upheld in a recent Michigan Supreme Court decision, which stated that the law effectively "give(s) every private entrepreneur and public agency a definite mandate to prevent pollution and environmental harm [and] imposes a duty on individuals and organizations both in the private and public sectors to prevent or minimize degradation of the environment which is caused or is likely to be caused by their activities" [*Ray vs Mason County Drain Commission,* No. 55248 (Sup. Ct. Mich. January 21, 1975)].

Since the act was passed in 1970, citizens or agencies have won more than 70 out of the 107 lawsuits filed against businesses or the governmental units that granted permits or licenses.

This act implies that any permit granted under the acts described in later sections can be contested in the courts if citizens or other agencies consider issuance of permits, or action to be taken thereunder, to be environmentally damaging.

As interpreted by the Michigan Supreme Court, the Michigan Environmental Protection Act prohibits pollution, destruction or impairment of the Environment, unless it can be shown that there is no feasible alternative, and that such pollution is consistent with the promotion of public health, safety and welfare in light of the paramount concern of the state for its natural resources. It appears from such decisions that the act gives the PSC standing to take legal action against other state agencies when these agencies attempt to promulgate or enforce environmental regulations that are not founded upon a proper balance of environmental and social welfare concerns.

Shorelands Protection and Management Act

The Shorelands Protection Act (Act No. 245, Public Acts of 1970) provides the basis for zoning and management of Great Lakes shorelines, particularly of critical natural areas and high-risk erosion areas. High-risk erosion areas have been identified on the Huron County shoreline of Saginaw Bay and along Lake Huron south of Harbor Beach in Sanilac County. Construction of an energy center, and especially a harbor to serve it, would require careful assessment. Permits are required to use the land for other than designated purposes.

Great Lakes Submerged Lands Act

The Submerged Land Acts (Act No. 247, Public Acts of 1955 as amended) governs filling and emplacing of structures, etc., for any purpose, on the bottom lands of the Great Lakes. Permits must be obtained from the Corps of Engineers, the State Waterways Commission and/or the local unit of government. This law implies that permits would be required to construct water intake and outlet systems across the bottom land area of Lake Huron for the energy center power plants. It would also have implications for construction of a harbor to serve the energy center.

Farmland and Open Space Preservation Act

This Act of 1974 (Public Act 116) was passed to help preserve farmlands and open space areas from nonfarm uses through property tax relief. Under the provisions of the act, farm and open space property owners may enter an agreement with the state whereby, in return for keeping this land in farm production or undeveloped, the owner receives a substantial tax credit. Each agreement lasts for not less than 10 years and can be renewed again for 10 years or longer.

At any time the landowner decides he no longer wants the agreement, he may apply to have the agreement relinquished. A lien is then placed against the land for the total amount of tax credit received plus an interest rate of six percent per annum compounded from the time credit was received until it was paid. At the time an agreement expires a lien is put on the property for the total amount of the tax credit received in the last seven years, and there is no interest payable. If the owner renews the agreement, the lien is discharged.

This act may have some implications on the cost of acquisition of land for an energy center. Property owners participating in this program will insist that sales profits must at least match, if not considerably exceed, the lien that they will be required to pay for withdrawing from the program. If a large acreage of Huron County farmland is enrolled in the program, it may be necessary to obtain a legislative mandate to release property owners from the agreements without the lien penalty in order to assure timely availability of the land for center purposes.

Soil Erosion and Sedimentation Act

The Soil Erosion and Sedimentation Control Act of 1972 (Public Act 347) requires that all earth change activities shall be conducted in such a way so as to prevent accelerated erosion and resultant sedimentation. Permits must be obtained from the designated county agency and are required for an earth change which affects any of the following land use activities on one or more acres of land:

1. transportation facilities
2. subdivisions or lot development

3. industrial and commercial development
4. service facilities
5. recreational facilities
6. utilities
7. oil, gas and mineral wells
8. water impoundment and waterway construction

This act cannot be construed as an obstacle to siting an energy center in Michigan, but it does require that erosion control measures be designed into plans for all phases of construction of the center and associated urban center. Noncompliance with the permits can be grounds for suspending construction activities.

Natural River Act

The Natural River Act of 1970 (Public Act 231) provides for establishment of a system of designated wild, scenic and recreational rivers in Michigan for the purpose of preserving and enhancing natural values. It is not anticipated that this act will have an effect on location of an energy center in the Harbor Beach or Montague areas of Michigan.

Wilderness and Natural Areas Act

The Wilderness and Natural Areas Act of 1972 (Public Act 241) was enacted to provide a mechanism for establishment and regulation of wilderness, wild and natural areas in the state of Michigan. To date, no areas for consideration under this act have been identified in Huron County. However, the Wilderness Advisory Board has discussed the desirability of designating at least one outstanding example of each landform/vegetation type in each of the natural regions of the state. The unique characteristics of the dunes in the Huron County area represent an outlyer of more northerly ecosystems. It may be desirable to designate the dunes a natural area, especially if they are found to support a colony of Butcher's thistle or other endangered plant species.

Inland Lakes and Streams Act

The Inland Lakes and Streams Act of 1972 (Public Act 346) was passed for the purpose of protecting the public trust in inland lakes and streams. It requires that permits be obtained to make any alteration to lakes or streams, including dredging, filling, putting structures on bottomlands, structural interference with natural flow, construction of artificial waterways and channelizing.

The act implies that permits will be required for many construction facets of an energy center plan, particularly in bridging and culverting streams for roads, haulways, etc. As with the Soil Erosion and Sedimentation Control Act, this act should not present a major obstacle, but represents another set of environmental constraints that will have to be satisfied, and noncompliance of which can cause delays or suspensions of activities.

Solid Waste Disposal Act

Under provisions of the act of 1965 (Public Act 87), as amended by Public Act 89 of 1971, Huron County has developed a solid waste management plan intended to serve the county until at least the year 2000. This plan is based on a relatively stable, predominantly rural population of about 30,000 people. Under this plan two sanitary landfill sites have been acquired. A 120-acre site in Sigel Township, near Bad Axe, was opened in November 1975. The second site includes 160 acres in Oliver Township. It is being prepared and should be ready for use soon.

With the advent of an energy center with coal-fired power stations and high waste-producing industries, a totally new solid waste disposal program would be required. Disposal of fly ash, to be expected from the coal-fired power plants, will require a much larger landfill area than now planned. Furthermore, depending upon the type of coal found and the mechanical processes used, the fly ash may be toxic. The existing landfill areas may not be suitable for disposal of toxic fly ash and potentially toxic wastes from the other associated industries. During the 20-year period of construction of the energy center, an urban center with a population of at least 100,000 will be evolving. This will require an estimated 850-1,000-acre additional landfill area, probably at the loss of agricultural acreage.

The problems of waste disposal from the 14 nuclear plants is another distinct problem still in need of a solution.

It is important that plans for development for an energy center in Michigan include plans to adopt "latest and best" solid waste disposal technologies as they develop in the future. The sanitary landfill solution must compete with agriculture for space. Alternative plans for waste disposal must be thoroughly evaluated to assure the most efficient and least environmentally damaging methods. The center should be designed, if at all possible, to utilize any otherwise vacant "buffer zones" between nuclear plants and other activities for solids waste disposal. It may be that landfills can be so designed that, once filled, the land can be returned to some type of agricultural production.

APPLICABILITY OF ENVIRONMENTAL STATUTES

In summary, there are a number of environmental statutes that will apply to any projected energy center in Michigan. None of them alone, except perhaps the Clean Air or Clean Water Acts, could stand as a decisive "go" or "no-go" factor. Approval of a permit by a state commission could be challenged under the MEPA if citizens felt significant environmental damage would occur. Furthermore, permits granted under any of the other statutes, such as the Inland Lakes and Streams or the Sedimentation and Erosion Control Acts, could be challenged.

The Michigan Department of Natural Resources (DNR) is the agency responsible for issuing most of the requisite permits (The Department of Agriculture and Department of Health also have responsibility for certain permits).

The DNR is in the process of streamlining the permit granting system and it may be that several permits could be applied for at the same time.

The magnitude of the energy center project is such that it may be more efficient to enact special legislation covering the entire project. This probably would require an extensive and detailed environmental impact statement covering all phases and facets of the total 20-year project.

FEDERAL CONSTITUTION AND STATUTES

U.S. Constitution, Article VI.
PL 92-532.
15 U.S.C. §§1 *et seq.* (1970).
15 U.S.C. §2a (1970).
15 U.S.C. §§12 *et seq.* (1970).
15 U.S.C. § 79 (1970).
16 U.S.C. §§ 791-828e (1970).
16 U.S.C. §§ 824 & 825 (1970).
16 U.S.C. § 824b (1970).
16 U.S.C. § 824d (1970).
16 U.S.C. § 824e (1970).
16 U.S.C. §§ 1451-1464 (1970).
33 U.S.C. § 403 (1970).
33 U.S.C. § 1151 (a).
33 U.S.C. §§ 1151 (b), 1160 (c) (3).
33 U.S.C. §§ 1151-75 (1970 ed Supp. II [1972]).
33 U.S.C. § 1251 (a).
33 U.S.C. §§ 1316 (a), (b).
33 U.S.C. § 1326 (b).
42 U.S.C. §§ 1857-58 (1970 ed. Supp. I [1971]).
42 U.S.C. § 1857 (a) (3) (1970).
42 U.S.C. § 1857 (b) (1) (1970).
42 U.S.C. § 1857c-3(a) (1970).
42 U.S.C. § 1857c-4 (1970). Michigan has submitted an implementation plan which has been approved by EPA.
42 U.S.C. § 1857c-5 (1970).
42 U.S.C. § 1857c-6(a) (2) (1970).
42 U.S.C. § 1857c-6(a) (4) (1970).
42 U.S.C. § 1857c-7 (1970).
42 U.S.C. §§ 4321-47 (1970).
42 U.S.C. §4332 (2) (c) (1970).

STATE STATUTES

Michigan Compiled Laws Annotated, § 336.15 (Supp. 1974).
Michigan Compiled Laws Annotated, § 323.5 (Supp. 1974).
Michigan Compiled Laws Annotated, § 460-6 (1967).

Michigan Compiled Laws Annotated, § 460.301 (1967).
Michigan Compiled Laws Annotated, § 460.501 (1967).
Michigan Compiled Laws Annotated, § 460.502 (1967).
Michigan Compiled Laws Annotated, §§ 460.551, 460.552, 460.555, 460.556 (1967).
Michigan Compiled Laws Annotated, § 460.557 (1967).

CASES

"Brooks v. Atomic Energy Commission, No. 72-2177 (D.C. Cir. March 8, 1973)", *Environ. Law Reporter,* 3:20219-20221 (1973).

Calvert Cliffs' Coordinating Committee v. US AEC, 449 F. 2d 1109 (DC Cir. 1971).

Citizens Organized to Defend the Environment v. Volpe, 353 F. Supp. 520 (S.D. Ohio 1972).

Connecticut Light & Power Co. v. FPC, 324 U.S. 515 (1945); hearings on Senate Bill 1925, 74th Congress, 1st Session (1935).

Environmental Defense Fund v. Corps of Engineers, 325 F. Supp. 728 (D.E.D. Ariz. 1971); *San Antonio Conservation Society et al. v. Texas Highway Dept.,* 446 F. 2d 1013 (5th Cir. 1971).

Green County Planning Board v. FPC, 455 f. 2d (2d Cir. 1973).

Hanley v. Mitchel, 460 f. 2d 640 (2d Cir. 1972).

Illinois Natural Gas Co. v. Central Illinois Public Service Co., 314 U.S. 498 (1942).

Julius v. City of Cedar Rapids, Iowa, 349 F. Supp. 88 (N.D. Iowa 1972).

Michigan Consolidated Gas Co. v. Sohio Petroleum Co., 321 Mich. 102, 32 N.W. 2d 353 (1948).

Ray v. Mason County Drain Commission, No. 55248 (Sup. Ct. Mich. January 21, 1975). *Environ. Law Reporter,* 5:20176-20180 (1975) (224 NW 2d 883).

Students Challenging Regulatory Agency Procedures v. U.S. 346 F. Supp. 189 (D.D.C. 1972).

United States v. Yellow Cab Company, 332 U.S. 218 (1947); *Jersey Central Power and Light Company v. FPC,* 319 U.S. 61 (1943).

"West Michigan Council v. AEC. No. G-58-73, (U.S. District Court for West Michigan)," *Environ. Law Reporter,* 5(11):310 (July 1974).

AGENCY RULINGS

Genesee Valley Gas Co., 3 SEC 672 (1938).
Monarch Mills, 1 SEC 822 (1936).
Pacific Gas & Electric Co., 10 SEC 39 (1941).
Texas Utilities Co., 31 SEC 367 (1950).
Wisconsin River Power Company et al., 27 SEC 539 (1948).

PUBLIC LAW

PL 93-633, 88 Stat. 2156.

REFERENCES

1. Hemdal, J. F., Ed. "A Preliminary Survey of Some Major Factors Bearing on the Feasibility of Energy Centers in Michigan," Draft for Review, Prepared by the Environmental Research Institute of Michigan for the Federal Energy Administration, FEA/G-77/060 (March 1977).

2. "Siting Energy Facilities at Camp Gruber, Oklahoma," Report to the Federal Energy Administration Prepared by the Pacific Northwest Laboratories of Battelle Memorial Institute, FEA/G-75/384 (June 1975).

3. "Energy Parks and the Commonwealth of Pennsylvania Issues and Recommendations," Prepared by the Center for the Study of Environmental Policy, The Pennsylvania State University, University Park, PA (July 1975).

4. "Puerto Rico Energy Center Study—Executive Summary," Prepared by Burns and Roe, Inc. and the Dow Chemical Company (May 1973).

5. "Industrial Complex Study for the Wasatch Front Area of the State of Utah," Prepared for Division of Water Resources, Utah Department of Natural Resources by the Dow Chemical Company (May 1973).

6. "Evaluation of Nuclear Energy Centers—Volume 2," Atomic Energy Commission, WASH-1288-VOL-2 (January 1974).

7. Toledo Edison Company. "Annual Report" (1975).

8. "Reliability and Adequacy of Electric Power 1974-1993," The Southwest Power Pool Coordination Council (April 1, 1974).

9. "Energy in Oklahoma," Oklahoma Energy Advisory Council (February 1, 1974).

10. Friedlander, Gordon. "Energy: Crisis and Challenge," Institute of Electrical and Electronics Engineers Spectrum (May 1973), pp. 21,26.

11. Duane, John W., and Michael A. Karnitz. *Power Eng.* (January 1975).

12. Moody, John D., and Robert E. Geiger. "Petroleum Resources: How Much Oil and Where?" *Technol. Rev.* (March/April 1975), p. 40.

13. "Assessment of Energy Parks vs Dispersed Electric Power Generating Facilities," Prepared by the Center for Energy Systems, General Electric Company for the National Science Foundation, Final Report (May 30, 1975).

14. Policy Study Group of the MIT Energy Laboratory. "Energy Self-Sufficiency: An Economic Evaluation," *Technol. Rev.* (May 1974), pp. 42,43.

15. "Proceedings of the First Meeting of the Advisory Committee on Energy Facility Siting—Volume II," Office of the Science Advisor, National Science Foundation, Washington, DC (November 1974).

16. "Energy Industrial Center Study," Prepared for the National Science Foundation by the Dow Chemical Company *et al.* (June 1975).

17. Brown, Charles Leonard. *Basic Thermodynamics* (New York: McGraw-Hill Book Company, 1951), pp. 135-138.

18. "A Preliminary Survey of Some Major Factors Bearing on the Feasibility of Energy Centers in Michigan—Executive Summary," The Environmental Research Institute of Michigan for the Federal Energy Administration (1977).

19. "Proceedings of the Workshop on Research Needs Related to Water for Energy," Research Report No. 93, Water Resources Center, University of Illinois at Urbana-Champaign, IL (November 1974).

20. Standard and Poor's *Industry Survey* (April 1975).

21. Gyftopoulous, E. P., L. J. Lazaridis and T. F. Widmer. *Potential Fuel Effectiveness Study,* Report to the Energy Policy Project of the Ford Foundation (Cambridge, MA: Ballinger Publishing Company, 1974).

22. "Basic Industrial Location Factors," Industrial Services No. 74, U.S. Department of Commerce, Washington, DC (June 1947).

23. "Environmental Considerations in Future Energy Growth," Battelle Columbus Laboratories and Battelle Pacific Northwest Laboratories (April 1973).

24. Bryan, R. H., and B. L. Nichols. "Summary of Legislative and Regulatory Activities Affecting the Environmental Quality of Nuclear Facilities," *Nucl. Safety* 13:264-274 (1972).

25. Boersma, L. K. A. "Integrated Systems for Utilizing Waste Heat from Steam Electric Plants," *J. Environ. Qual.* 2:179-187 (1973).

26. Mount, P. I. "Developing Thermal Requirements for Freshwater Fishes," in *Biological Aspects of Thermal Pollution, Proceedings National Symposium* (Portland, OR: Vanderbilt University Press, 1968), p. 140.

27. "Why Catfish?" Excerpts from several unpublished reports of the Michigan State University Department of Agricultural Engineering.

28. Strawn, K. "Beneficial Uses of Warm Water Discharges in Surface Waters," in *Electric Power and Thermal Discharges* (New York: Gordon & Breach, Science Publishers, 1970), p. 143.

29. Parker, F. L., and P. A. Krenkel. "Thermal Pollution: Status of the Art," National Center for Research and Training in the Hydrologic and Hydraulic Aspects of Water Pollution Control, Report No. 3. Vanderbilt University, Nashville, TN (1969).

30. Mihursky, J. A. "On Possible Constructive Uses of Thermal Additions to Estuaries," *Bioscience* 17:698 (1967).

31. "Agricultural Uses of Waste Heat," Prepared by Michigan State University, Agricultural Engineering Department for the Environmental Research Institute of Michigan (October 31, 1975).

32. Agardy, F. J. "Waste Heat Utilization in Wastewater Treatment," PB 217880 (January 1973).

33. "Waste Heat Utilization from Power Plants in Agriculture," Agricultural Engineering Department, Michigan State University (June 1975).

34. Brooks, Richard Oliver. "New Towns and Communal Values," in *A Case Study of Columbia, Maryland* (New York: Praeger Publishers, 1974).

35. Gipe, Albert B. "Planning a New City—Columbia," *The Transactions of Industry and General Applications, IEEE*, Vol. IGA-2No5 P 423-430 (September/October 1966).

36. Miller, A. J., *et al.* "Use of Steam-Electric Power Plants to Provide Thermal Energy to Urban Areas," Report No. ORNL-HUD-14, Oak Ridge National Laboratory (January 1971).

37. "Environmental Report, Quanicassee Plant Units 1 and 2," Consumers Power Company, Jackson, MI.

38. *Michigan Statistical Abstracts* (Lansing, MI: Michigan State University, 1974).

39. National Park Service. "National Register of Historic Places," *Federal Register* 38(39) (February 28, 1973).

40. Shaker, W. H., and W. Steffy, Eds. "The Consumer Speaks Out on Electric Energy Policy," in *Consumer's Guide to Electric Power Reform: The Alternatives for Michigan,* Institute of Science and Technology, Industrial Development Division (Ann Arbor, MI: The University of Michigan, 1975).

41. "Preliminary Investigational Requirements—Petrochemical and Refinery Waste Treatment Facilities," Water Pollution Control Research Series No. 12020 EID (March 1971).

42. "Water Quality Management Planning for Urban Runoff," U.S. Environmental Protection Agency, Report No. EPA-440/9-74-004 (December 1974).

43. Sartor, J. D., and G. B. Boyd. "Water Pollution Aspects of Street Surface Contaminants," U.S. Environmental Protection Agency, Research and Monitoring Environmental Protection Technology Series, EPA-R2-72-081 (November 1972).

44. *Great Lakes Basin Framework Study* (Ann Arbor, MI: Great Lakes Basin Commission, 1975).

45. Weinberg, A. M., and R. P. Hammond. "Limits to the Use of Energy," *Am. Sci.* 58(4):412-418 (1970).
46. Hanna, S. R., and S. D. Swisher. "Meteorological Effects of the Heat and Moisture Produced by Man," *Nucl. Safety* 12(2):114-122 (1971).
47. Rotty, R. M. "Atmospheric Thermal Pollution," paper presented at the Conference and Workshop of the American Meteorological Society, Asheville, NC, October 8, 1974.
48. Joint Committee on Atomic Energy. "Selected Materials on Environmental Effects of Producing Electric Power," 91st Congress, First Session, U.S. Government Printing Office (August 1969).
49. *Recommendations of International Committee on Radiation Protection* (ICRP), Publication 6 (London: Pergamon Press, 1969).
50. Rasmussen, J. "Reactor Safety Study, An Assessment of Accident Risks in U.S. Commercial Nuclear Power Plants," WASH-1400, U.S. Nuclear Regulatory Commission (1975).
51. Cuddihy, R., *et al.* "A New 'Clean Image' for Coal," *Environ. Sci. Technol.* 11(13):1148-1165 (December 1977).
52. Kaakinen, J. W., R. M. Jorden, M. H. Lawasani and R. E. West. "Trace Element Behavior in Coal-Fired Power Plant," *Environ. Sci. Technol.* 9(9):862-869 (September 1975).

GLOSSARY

AC—*Abbreviation for* alternating current. In this country, the current alternates direction 120 times per second, or 60 cycles per second.

Benthic Organisms—Organisms living on or at the bottom of a body of water.

By-Product Power Generation—Similar to cogeneration but usually signifies that steam is the major product and electric power is a by-product.

Coalplex—An energy center with only fossil-fueled power generation and possibly coal gasification and/or liquefaction.

Cogeneration—See "dual-purpose."

Collocation—The construction and operation of electric power generation facilities and industrial users of power, steam and heat, in close proximity for their mutual benefit.

Curie—The basic unit to describe the intensity of radioactivity in a sample of material. One curie equals 37×10^9 disintegrations per second, or the radioactivity of one gram of radium.

DC—Direct current. Current always goes in the same direction.

Dual-Purpose Central Power Stations—By "dual-purpose" is meant a central power station, of size and type typically constructed by the utilities, which also furnishes a significant amount of steam to one or more customers.

Energy Park—Energy center.

Enthalpy—The sum of the internal energy and a pressure-volume product of a fluid. This sum is commonly used to simplify energy calculations. The internal energy is that stored within a mass of molecules by reason of their random motion.

HVDC—High-voltage direct current. For the transmission of electric power, the voltage may be several hundred kilovolts.

MBtu—Millions of British Thermal Units. A Btu is the heat required to raise one pound of water 1°F.

mCi—Milli curie: 1/1,000 of a curie.

mil—Used to describe the cost of one kilowatt-hour of electricity in tenths of one cent. 30 mils is equivalent to 3¢/kWh.

Nektonic Organisms—Organisms in a body of water swimming independently or capable of moving against the flow.

Noncoincident Peak Load—The combined peak load of two utilities at any one time. It is not the same as the sum of the separate peak loads of each utility since their peaks usually occur at different times.

Nuplex—An energy center with nuclear-fueled power generation.

Planktonic Organisms—Passively floating or drifting organisms in a body of water.

psia—A pressure unit in pounds per square inch absolute or the pressure above zero.

psig—The pressure in pounds per square inch above atmospheric pressure.

PVC—Polyvinylchloride, a plastic commonly used for the manufacture of pipe.

rad—One rad is equal to the absorption of 100 ergs of radiation energy per gram of matter.

rem—A unit of absorbed radiation dose in biological matter. A dose in rems equals the absorbed dose in rads multiplied by the relative biological effectiveness of the radiation.

Secchi Disk—A flat disk painted black and white in alternate quadrants. When lowered into a body of water, the depth at which the disk is just discernable is a measure of the clarity or turbidity of the water.